WALL OF STEEL

3·7 in action in North Africa.

(Fergal Kearns)

WALL OF STEEL

The History of the
9th (Londonderry) Heavy Anti-Aircraft Regiment
Royal Artillery (Supplementary Reserve)

RICHARD DOHERTY

NORTH-WEST BOOKS

1988

First published in 1988 by
North-West Books
23 Main Street, Limavady, County Londonderry,
Northern Ireland

BRITISH LIBRARY CATALOGUING IN PUBLICATION DATA

Doherty, Richard
Wall of Steel: the history of the 9th (Londonderry)
Heavy Anti-Aircraft Regiment, Royal Artillery
(Supplementary Reserve).
1. World War 2. Army operations by Great Britain.
Army. Royal Artillery (Supplementary Reserve).
Heavy Anti-Aircraft Regiment, 9th (Londonderry)
940.54'12'41

ISBN 0-907528-13-9

Printed by
Graham & Sons (Printers) Ltd., 51 Gortin Road, Omagh, Co. Tyrone, N. Ireland

DEDICATION

For my father, Gunner J J Doherty,
and all of his comrades who served in the "Ninth"

Once a gunner, always a gunner

CONTENTS

CONTENTS—continued

ILLUSTRATIONS

ILLUSTRATIONS—continued

MAPS

ACKNOWLEDGEMENTS

Producing this record of 9th (Londonderry) Heavy Anti-Aircraft Regiment would have been impossible without the willing help of a large number of organisations and individuals who gave me tremendous assistance in my quest for information. I appreciate that help very much and wish to place on record my thanks to everyone involved. Of course the final writing of "Wall of Steel" fell to myself and any errors which may have occurred in the work can be laid at my door. However I hope that the help and advice which I have received will have ensured that the account is as accurate as possible and I therefore gratefully acknowledge the aid and co-operation of:

PUBLIC RECORD OFFICE, Kew, Richmond, Surrey (especially the staff of the Search and Reading Rooms for their patience and helpfulness)

ROYAL ARTILLERY INSTITUTION, Woolwich, London (especially Brigadier R J Lewendon, RA (Retd), Historical Secretary and Mr G Sutton, ALA, Librarian)

IMPERIAL WAR MUSEUM, Lambeth, London (especially the Department of Photographs and the Department of Printed Books)

NATIONAL ARMY MUSEUM, Royal Hospital Road, London

ROYAL ARTILLERY ASSOCIATION — LONDONDERRY BRANCH

ROYAL ARTILLERY ASSOCIATION — LIMAVADY BRANCH

ROYAL ARTILLERY ASSOCIATION — BALLYMENA BRANCH

6 LIGHT ANTI-AIRCRAFT BATTERY OLD COMRADES' ASSOCIATION, Coleraine

ROYAL BRITISH LEGION, Newtownards

PUBLIC RECORD OFFICE NORTHERN IRELAND, Balmoral Avenue, Belfast (especially Dr A W P Malcolmson)

COMMONWEALTH WAR GRAVES COMMISSION, Maidenhead, Berkshire

THE QUEEN'S ROYAL SURREY REGIMENTAL ASSOCIATION, Canterbury

In addition to the organisations listed above my thanks are due also to:

Colonel Ian B Gailey, TD, whose knowledge of the reserve and volunteer forces in Northern Ireland was invaluable.

Mr James Lucas, Detling, Kent for his advice and guidance and his kind permission to draw upon his excellent book "War in the Desert — The Eighth Army at El Alamein".

Sir John T McFarland, Bt TD for his permission to inspect his late father's papers, for the use of photographs from his father's collection and for additional information on Sir Basil McFarland.

Sister Mary Richard Coyle, OFM, Missionary Franciscan Sisters, Rocci, Rome for her help on Appendix 2 and for information on wartime Alexandria.

Most of all I owe a great debt to the many members of 9th (Londonderry) Regiment who have shared with me their memories, given freely of their advice and permitted access to diaries or journals and photographs. If I have succeeded in capturing the human aspects of the Regiment's history it is due entirely to the patient, good-humoured gunners who helped me. They are:

Messrs R Hamilton, A W Hogg, W Hunter, T Moore, M Mulhern, G McCallion, T McCready, Jack McIlmoyle, Joe McIlmoyle, T Patton, J Russell, J Taylor, J Wilson (all of 24 Battery)

E H Babington, J Canning, H Cuthbert, J Dinsmore, G Lapsley, A Missen, J McMahon, J J Ormsby, J Peoples (deceased), W Sherrard, W Smyth, J Soutar, M Stott, A Taylor (deceased), E Wilson (all of 25 Battery)

ACKNOWLEDGEMENTS—continued

W Black, W Darragh, E Fallis, T Hinds, G Kernoghan, R Montgomery, T Moore, A McClean, J McConnell, D McCurdy, C McCluggage, A McNinch, J Neilly, T Nixon, H Wilson (all of 26 Battery)

J Atcheson, J Dorrian, E Griffiths, J Mawhinney, J McGinn, H Paterson, J Pritchard (all of 5 LAA Battery)

S Anderson, R Balmer, R Burns, D J Christie, R Gage, R McDonald, T McDonald, D Pringle, W Oliver (all of 6 LAA Battery)

I regard it as a singular tribute to the Regiment that Major General B P Hughes, CB CBE, agreed to write a foreword to this book. A Senior Historian of The Royal Regiment of Artillery, General Hughes is also a former Colonel Commandant of the Royal Regiment and was one of the most distinguished of wartime gunner officers. I thank him for reading the manuscript and providing a fine foreword.

Mr John Conacher of North-West Books must be given a special word of thanks for agreeing to publish the book and for his enthusiastic support of the project as it developed as well as his valuable expert advice.

The inclusion of photographs was made possible by the assistance of a number of sources. Each photograph is attributed to its source in the text but I must thank the Imperial War Museum who have kindly permitted the reproduction of nine excellent pictures. Closer to home the collection of Terence McDonald and David Bigger proved invaluable as did those of Sir Basil McFarland and a number of other 9th Regiment members especially John Taylor, John Ormsby and Jim Canning as well as Tommy McDonald of 6 LAA Battery. Some interesting photographs were rather worn snapshots but the professionalism of Miss Andrea O'Hare ensured that they were copied and enlarged while remaining suitable for reproduction in this book, where they have been further enhanced by the care and skill of the printers.

Maps for the book were drawn by Mr Fergal Kearns who also produced the excellent drawing of a 3·7 inch HAA gun in action which is used as the frontispiece.

Major Sam Irwin, RA (TA), Battery Commander of 206 (Coleraine) Air Defence Battery was an enthusiastic source of support and assistance. His Battery, descended from 9th and its sister units, very obviously has inherited the spirit of those older gunners.

Along the way a number of people have helped to point me in the right direction from time to time. For that help I thank Colonel James Hughes, CBE TD, President of the Royal British Legion in Northern Ireland, Major Roy Graham, TD, Chairman of the Northern Ireland Region of the Royal Artillery Association, my very good friend Joe Sandison, QGM, who was an especial and indispensible help on more than one occasion and Mr John Dinsmore, Editor of the North-West edition of the Belfast Telegraph.

The text of "Wall of Steel" was typed — and retyped — by Mrs Anne Griffiths who did a first-class job while I cannot fail to mention another lady — my wife Carol — without whose patient support and tolerance over a sustained period this book could never have been written.

To all of those I have mentioned — and to anyone I may have forgotten — I extend my deepest gratitude.

Richard Doherty

LIST OF CONTRIBUTORS

The author and publisher express their warm thanks to the undernoted, whose help in making possible the publication of this book is greatly appreciated.

The Honourable The Irish Society
Limavady District Council
Coleraine Borough Council
Strabane District Council
Royal Artillery Association — Londonderry Branch
Royal Artillery Association — Limavady Branch
Royal Artillery Association — Ballymena Branch
The Services Club, Londonderry
The Royal British Legion — Northern Ireland Area (Number 10 Branch)
The Royal British Legion — Enniskillen Branch
The Royal British Legion — Ballymoney Branch
The Royal British Legion — Portrush Branch
Mr James Canning, Londonderry
Mr Bertie Cuthbert, Londonderry
Mr George Lapsley, St Johnston, Co Donegal
Mr Ernest McClure, Londonderry
Mr Donald McBurney, Belfast
Mr J J Ormsby, Strabane
Mr William Smyth, Strabane
Mr Eric A. Wilson, Strabane
Mr Jack Wilson, Newtownabbey
102 (Ulster) Air Defence Regiment, Royal Artillery (Volunteers)
206 (Coleraine) Air Defence Battery, Royal Artillery (Volunteers)
D Coy, 7th/10th (City of Belfast) Bn., The Ulster Defence Regiment
The Territorial, Auxiliary and Volunteer Reserve Association for Northern Ireland
Northern Bank Ltd., Shipquay Place, Londonderry
Bank of Ireland Ltd., Shipquay Street, Londonderry
John Kelly Ltd., (Coal Importers), Bay Road, Londonderry
T J McConnell & Sons Ltd., Springtown, Londonderry
Atcheson & Son (Derry) Ltd., Glenaden, Londonderry
The White Horse Inn Ltd., Campsie, Londonderry

FOREWORD

By Major General B P Hughes, CB CBE
A Senior Historian of the Royal Regiment of Artillery

It gave me great pleasure to have been asked to write a foreword to this history of the 9th (Londonderry) Heavy Anti-Aircraft Regiment, Royal Artillery (SR).

In the first place that is because, while I was serving both during and after the Second World War, there were four separate occasions during which I was in contact with this unit, and therefore know a great deal about its magnificent spirit, and its achievements in so many different theatres of war. It is indeed good to be reminded in the text of several old friends such as Frankie Dearden, 'Bob' Aitken and Jim Longfield.

There is however, another pleasure in reading such an excellent and well compiled history. While it will clearly be of great interest to the old members of the unit itself, and will bring back many good memories to them, there is an even greater merit to be acclaimed. A meticulously compiled history such as this, containing the first hand evidence of so many participants, will be of the greatest value to those of our successors who will be concerned with military history in the future. Many details of persons, times and places are so easily forgotten, but here is a reliable record for all time.

Finally may I congratulate the compiler who has put together such an excellent record of this unit's fine war service.

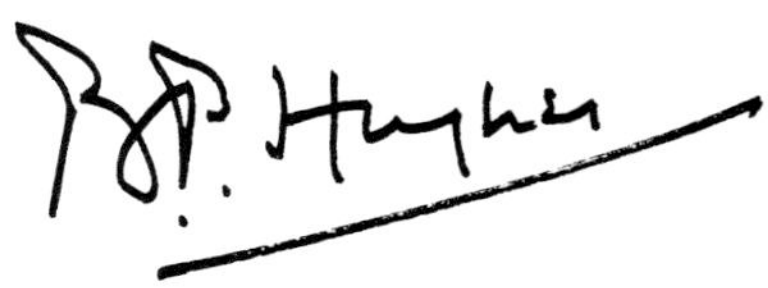

INTRODUCTION

Invariably the last piece of a book to be written is the introduction, and "WALL OF STEEL" is no exception to this rule. For more than a year the crafting of this book has loomed large in my life — indeed its creation has been something towards which I have been working for some fifteen years.

I belong to the first generation to be born after the Second World War — into a world that was still gradually trying to come to terms with the peace that had been achieved and the tremendous price that had been paid for that peace. During the sixties youth throughout the world clamoured for lasting peace; it became almost fashionable to question the motives of the men who had fought the 1939-45 war and gained the peace which had saved my generation of young Europeans from being offered as a blood sacrifice to the god of war — as had been the fate of our fathers and grandfathers. There were times when the younger generation scorned their immediate predecessors and indeed blamed them for the existence of what President John F Kennedy once described as "a nuclear sword of Damocles".

Since then attitudes have changed considerably and there has been a tremendous upsurge in interest in the Second World War and in the men who fought in it. Much of that interest has been centred around what might be described as the "high profile" activities of the combatants, and little popular interest has been taken in aspects such as the anti-aircraft artillery with which this book is wholly concerned. In my own reading I have often been surprised to see how little attention has been paid to the gunners generally but especially to the men — and women — who manned the heavy "ack-ack" guns. That dearth of attention is all the more surprising when one considers that a quarter of all members of the wartime army served in the Royal Regiment of Artillery. Since the gunners were everywhere, it seems that they were quite simply taken for granted.

The lack of attention given to the work of the "ack-ack" gunners was one of the stimulants which prompted me to write this book. Another was the fact that there has never been a time in my life when I was not aware of the existence during the Second World War of a Heavy Anti-Aircraft Regiment of the Royal Artillery which had been raised largely in my native city and indeed took its title of Londonderry from the city. Of the Regiment's service and achievements I knew little. Although he served in the Regiment my late father was not a man to describe the horrors of war to anyone — especially to his children, although we did hear of many humorous incidents and learned a lot about the many countries in which he served in war and peace. However it was very obvious that he was proud of his gunner regiment, as he was a regular attender at reunion functions and never missed a Remembrance Sunday parade while he was fit enough to attend. No doubt that quiet, unspoken pride in the "Derry Boys" which he demonstrated was another — and perhaps the most important — factor in spurring me to learn and to write down the story of what I hope I will be excused of describing as "our" Regiment.

The story which I was to discover through many hours of research and interviews was a fascinating one. This story does great credit to the men who served in 9th HAA Regiment, for they did their duty extremely well and have, in the years since 1945, almost without exception been equally modest about their war service.

Although the Regiment was a new one in a relatively new role the tradition into which the Regiment fitted was a long and honourable one. Derrymen fought in the scarlet coats of Wellington's army in the Peninsula and at Waterloo, where, according to the Iron Duke himself, the stubborn bravery of the 27th (Inniskilling) Regiment saved the centre of the line. Two battalions of that Regiment, by then renamed Royal Inniskilling Fusiliers, served in the South African War. During the Great War from 1914 to 1918 a Derry Battalion — the 10th (Derry Volunteers) — of the Inniskillings served as part of 36th (Ulster) Division. The infantry tradition was therefore strong in the city and district, but the gunner tradition was by no means new. In 1881 the old Londonderry Light Infantry Militia (formed 1793) had been amalgamated with the Royal Tyrone Fusiliers Militia to form the 4th Battalion of the newly created Royal Inniskilling Fusiliers. However, soon after, in April 1882, the Londonderry element of the 4th Battalion was reorganised as a Militia Artillery unit in the North Irish Division. This new Regiment — or Brigade, which was the term used in the Royal Regiment until 1938 — was numbered as 9th Brigade giving a fortuitous connection with the new Regiment formed when the gunner tradition restarted in the city in 1939. (There had been an earlier Londonderry Artillery Militia unit from 1855 to 1875 when it was amalgamated into the Mid-Ulster Artillery Militia. 9th Brigade became the Londonderry Artillery (Southern Division, RA) and then the Londonderry Royal Garrison Artillery (Militia) before becoming part of the Special Reserve Royal Field Artillery in 1908; it was disbanded in 1909.)

For the volunteers of 1939 there was the very real threat of war which most of them believed to be close at hand. They were also aware that, as a Supplementary Reserve unit, their Regiment would be joining the Army's field forces, with France and the British Expeditionary Force their most likely destination. There was however none of the "jingoism" which had marked the beginning of the 1914-18 War, especially as few of the men came from families which had not been touched by the terrible toll of the earlier war.

It was six years before their job was accomplished, and during those years the Regiment served with distinction in North Africa and Italy, with many of its members losing their lives. That 9th HAA Regiment was able to achieve the high level of efficiency, professionalism and respect which they did was due to a number of factors. Their initial permanent cadre of officers, warrant officers and NCOs included some excellent and inspiring men among whom were numbered Stanley Smith, the first Commanding Officer, and his Battery Commanders Walter Lambert, Frank Dearden, Tony Traill, William Allison and Karl Webb. They paved the way for their successors handing over a fine Regiment to Smith's replacement, "Bob" Aitken, a man who was universally respected and admired in the Regiment. Aitken, in the words of Sir Basil McFarland "made the Regiment". To do so he had the help of such excellent Battery Commanders as Sir Basil himself, Jimmy Henderson of 24 HAA Battery from Co Fermanagh, and Sam Haughton of 26 HAA Battery from Ballymena — all of them Supplementary Reserve officers. The quality of the original officers surely gives the lie to the oft-repeated falsehood that the officers of the 1920s and 1930s were not professionals — how could leaders such as Montgomery,

Alexander, Wavell and Slim have arisen from an unprofessional body? And these are only four from scores of names which could be listed.

Such were the men who turned civilians into gunners of the highest quality but the civilians themselves had, by and large, been an excellent grouping. Speaking of them almost fifty years later E H (Sonny) Babington proudly describes them as "Derry's best". With men like Walter Wray, Jim Canning, Freddie and Jim Peoples (both of whom unfortunately died while this book was in preparation), Tommy McCready, Jimmy McFeely and Neill Murray, to name but a few, there is little doubt that his claim is a just one.

It was a tremendous advantage that the Regiment's personnel already had local loyalties and common ties, whether those were in Londonderry, Ballymena, Coleraine or any of the other towns and districts which were represented in their ranks. In addition they quickly absorbed many of the traditions of the Royal Regiment of Artillery and while so doing created their own very special identity as the "Derry Boys". It would be pertinent to add that there were further special identities, as they soon discovered that the first loyalty of any gunner is to his Battery. Even today the friendly rivalry of the individual Batteries is apparent at any gathering of gunners of the 9th. "Two-Four", "Two-Five" and "Two-Six" Heavy AA and "Five" and "Six" Light AA Batteries are as strong in spirit today as they were all those years ago.

It is that spirit which I have tried to capture in this book, and I have done my best to give a fair balance to each of the Batteries, showing the strengths and achievements of all of them. For me the writing of "Wall of Steel" has been a pleasurable task. So too the gathering of information for it, which quite literally was a labour of love. I received the most enthusiastic co-operation everywhere I went and, on a personal note, it has been both a touching and a humbling experience to see the esteem in which my late father was held by his comrades. The very mention of the fact that "I am J J's son" opened doors to many new friendships.

My hope is that this book will provide a tangible tribute to all of the men who served in 9th (Londonderry) HAA Regiment — not just the men from the Regiment's home city but those from all over Northern Ireland, from England, Scotland, Wales and the Channel Islands, from Eire (as it then was) and from far-off Basutoland in southern Africa. I do not claim that they were any form of supermen or heroes — nor would they wish me to do so; what I do claim is that they were ordinary men from many different walks of life who did their wartime job extraordinarily well, coping with their own fears as when they came under intensive shellfire on the Salerno beachhead, and stoically accepting hardships and their long exile from home. (It is unlikely that any other unit spent so long overseas.) They have the satisfaction of knowing that they played their part in bringing about the defeat of Naziism in a war that has been described by the historian A J P Taylor and by Pope Pius XII as a just war. And the "Derry Boys" were on the side of justice.

Richard Doherty

Action Stations! A night raid on Alexandria Harbour, 1940.

(Photo: John Taylor)

PART ONE

THE MAKING OF A REGIMENT

CHAPTER ONE

THE BEGINNINGS

On January 30th 1933, Adolf Hitler became Chancellor of Germany. Although he had come to power by using democracy this soon began to wither in the new Germany. Four weeks later the *Reichstag*, the German Parliament, suspended civil liberties and on March 23rd passed the Enabling Act which gave Hitler absolute power. Germany left the League of Nations in October 1933, and the following month elections were held in which Hitler's National Socialists took 92% of the vote. When President Hindenburg died in August 1934 Hitler became *Führer*, or Leader, of the German Reich and assumed supreme command of the German forces.

The expansion of the army and navy and the establishment of an air force, the *Luftwaffe*, which Hitler ordered in October 1934, directly contravened the Treaty of Versailles of 1919. In attempting to curb, and indeed actively prevent, the regrowth of Germany as a military power the Versailles settlement, in which Germany had no part, had imposed such conditions on the vanquished as to help provoke the very thing that the Treaty allegedly set out to stop — the resurgence of the Prussian military machine.

Other European nations began to awaken to the danger created by Hitler. In the United Kingdom however calls for re-armament had been seen as, at best, scaremongering, and, at worst, as warmongering. The war-weariness following 1918, the memory of the awful slaughter of those four years and the economic problems of recent years all contrived to produce an attitude of isolationism. Even so a number of expansion schemes were drafted and began to be put into effect but the unwillingness to be embroiled in another European war was exemplified by the reluctance to create an Expeditionary Force for use in Europe. The memory of the Somme had not diminished — nor it seemed had the memory of Waterloo, for ironically some of the defence plans were drafted around the possibility of war with France! The assumption of an enemy just across the Channel was of course to become a reality when Germany overran and occupied France.

When Britain finally awoke to the real menace of Hitler it was almost too late. Only after Chamberlain had travelled to Munich to meet Hitler and returned with the famous piece of paper guaranteeing "peace in our time" did the Cabinet approve the establishment of a British Expeditionary Force for use in Europe. That decision was taken on February 22nd 1939 and three weeks later Germany tore up the Munich Agreement.

Until now Britain's primary concern had been with defence of the United Kingdom itself. One of the main factors in this concern was the threat of air attack since it was widely believed that the bombers would always get through any fighter defences. To guard against bombers a scheme for anti-aircraft artillery defence was drawn up requiring over 43,000 men. This entailed converting many units of the Territorial Army to Royal Artillery — beginning with some of the London Territorials — and it led to the creation on December 15th 1935 of 1st Anti-Aircraft Division to defend London. By 1939 an Anti-Aircraft Command of seven divisions had been set up, although its equipment came nowhere near to matching its manpower. (The plans called for a force of nearly 1,300 heavy guns, 2,000 light guns and almost 5,000 searchlights, yet only 50 of the new 3·7 inch guns were ready and not even one of the heavier 4·5 inch guns had been made available to the Royal Artillery).

Of course the BEF would need anti-aircraft defences also as would British possessions and overseas bases. All these commitments had to be added to the home requirements for anti-aircraft artillery. Thus it is hardly surprising that arguably the greatest expansion in the army's strength came in the relatively new art of anti-aircraft gunnery. Most of this expansion occurred in the Territorial Army which assumed responsibility for the "gun and searchlight defence of the country against hostile air attack".

A lesser-known element of the army's reserves also took part in this expansion — the Supplementary Reserve.[1] (SR). Although normally administered by the Territorial Army Associations, the Supplementary Reserve was part of the First Class Army Reserve liable for service anywhere in the world and for call-out whenever the Army Reserve or any part of it was called out by Proclamation. This allowed the Government to call out units of the Supplementary Reserve in advance of a declaration of war.

In July 1938 the first public announcement of the creation of a Northern Ireland-recruited Anti-Aircraft Brigade of the Supplementary Reserve was made. Recruiting officially began on April 3rd 1939, coincidentally only days after the Secretary of State for War, Leslie Hore-Belisha, announced plans to double the TA. 3rd (Ulster) Anti-Aircraft Brigade (SR) began to recruit throughout the Province on that day for its three regiments and ancillary support units. These were:

8th Anti-Aircraft Regiment, RA (SR)
9th Anti-Aircraft Regiment, RA (SR)
3rd Searchlight Regiment, RA (SR)
3rd AA Brigade Signal Company, Royal Corps of Signals (SR)
3rd AA Brigade Company, Royal Army Service Corps (SR)
3rd AA Brigade Workshop Company, Royal Army Ordnance Corps (SR)

9th Regiment was to be based in Londonderry and it soon adopted, unofficially, the

[1] The Supplementary Reserve was descended from the old Militia which had ceased to exist on the formation of the Territorial Force in Great Britain in 1908. Some Militia units survived however being renamed Special Reserve. After the First World War the Special Reserve was renamed Supplementary Reserve.

title of 9th (Londonderry) Anti-Aircraft Regiment. It was not in fact until May 1940 that the War Office approved the use of Londonderry in the title.

As with 8th Regiment the Derry unit comprised three heavy and one light batteries. These were 24, 25 and 26 AA Batteries and 6 Light AA Battery. Regimental Headquarters together with 24 and 25 Batteries were to be based in the City — initially at Ebrington — while 26 Battery was at the Demesne, Ballymena (later to become Lowfield Camp) and 6 Light AA Battery at Calf Lane, Coleraine. Thus the Regiment could recruit across the north of the Province from Londonderry to Larne.[2] However, that did not mark the actual boundary of its recruiting territory since it also took men from around Strabane and Sion Mills in Co Tyrone and from the Fermanagh area.

Each of the batteries had detachments removed from its own locations. 24 Battery had men from Limavady and Fermanagh as well as from Londonderry while the Co Tyrone detachment of over twenty men belonged to 25 Battery. 26 Battery had detachments at Larne, Bushmills and Castledawson. In addition there were a number of men from neutral Eire who joined up using addresses in Northern Ireland. Most of these were Donegal men but at least one came from as far away as Co Mayo; Paddy Lawlor, an ex-infantryman, walked and hitch-hiked all the way from his home town to join the Regiment when he heard that Major Walter Lambert, under whom he had served in India, had been appointed to command an anti-aircraft battery.

However, before any men enlisted the recruitment and selection of officers had already taken place. The first officers for the Regiment came from either the Regular Army or the half-pay list and were appointed from January 1st 1939. Command of the Regiment was vested in Lieutenant-Colonel G L Stanley Smith, OBE, RA, who, with his adjutant, Captain J G Tyrrell, arrived at Ebrington on January 15th 1939. They had been preceded by Battery Quarter-Master-Sergeant (BQMS) "Buck" Jones of 25 Battery who had arrived on January 1st, the official "date of birth" of the Regiment.

The Colonelcy of the Regiment, an honorary position, was offered to and accepted by the Governor of Northern Ireland, His Grace the Duke of Abercorn.

Major Walter Lambert, an ex-Connaught Rangers officer, was gazetted to command 24 Battery with Major Frank Dearden in charge of 25 Battery. Across at Ballymena Major Anthony O'Brien Traill commanded 26 Battery and Major William W "Toby" Allison was the man responsible for 6 LAA Battery at Coleraine.

At the time it was proposed that the nucleus of regulars — officers and NCOs — would remain with the Regiment for two to three years to assist in its formation, but events were to overtake these ideas. As the regulars arrived a Regimental Headquarters (RHQ) and Battery offices were established in wooden huts in Ebrington Barracks and it was here that the candidates for Supplementary Reserve commissions presented themselves for interviews and medical tests in January 1939.

[2] The recruiting areas for 3rd AA Brigade were decided on as the counties of Antrim, Down and Londonderry with the cities of Belfast and Londonderry. The North Irish Horse (SR) were to recruit in the remaining counties. However as can be seen in this Chapter such areas were not watertight.

The local officers came from a variety of walks of life. There were businessmen, solicitors and teachers and many of their names were well-known in the business and professional life of the City. Most of the SR officers were commissioned from April 1st but some held commissions dated from February 20th. Among the latter were Sir Basil McFarland, Bt., the second Baronet Aberfoyle who was soon to be elected Mayor of Londonderry, A J Welch from a well-known local business family, and Sam Haughton from Cullybackey, all of whom were to command batteries in the Regiment. Over forty candidates were accepted for commissions in the four batteries with some two dozen of these being in 24 and 25 Batteries in Londonderry.

A Regimental Information Bureau was also set up in a stockroom of the Northern Counties Hotel in the centre of Londonderry. This was placed at the Regiment's disposal by the hotel manager, Mr G S Maxwell, who had been commissioned into 24 Battery. When recruiting officially commenced in April this Bureau became the recruiting office for the Regiment. Through the offices of Mr Maxwell an Officers' Mess was also established in the Northern Counties.

The initial nucleus elements of the Regiment spent their time preparing for the influx of recruits after April and this included ordering equipment and kit and updating training manuals. However the recruiting, when it did begin, was so brisk that the supply services took some time to catch up with the Regiment's needs. In fact by April 14th 24 Battery could already boast some 100 men and there was a similar picture in each of the other batteries.

As the recruits came in, plans were already in hand to move them out of Ebrington; it was home to an infantry battalion and could hardly accommodate another regiment as well. The new home for the gunners was under construction further along Limavady Road at Caw. On May 8th 1939 the Regiment began its move into Caw Camp.

At that time Caw, which was a camp of wooden hutments, was incomplete and indeed a fence was still being constructed around it as the permanent staff began to move in. One of the men who had been building the camp was an ex-gunner called Matt Mulhern. Mulhern was working at the fence when Sergeant "Lofty" Muir arrived. The two had served together before and Muir soon talked his old mate into enlisting. As Mulhern had previous experience in the Royal Artillery he was promoted immediately to Lance-Sergeant.

The first anti-aircraft equipment arrived on April 29th just before the move from Ebrington to Caw when two 3-inch guns, a Predictor and a Height and Range Finder were delivered to 25 Battery, who were the Headquarters Battery. No live rounds were supplied for the guns and training had to be carried out with wooden rounds!

Training programmes were under way for each of the four batteries, the men from the outlying detachments being lorried or bussed in to their respective batteries for training nights. In addition to training on the guns and instruments there was the inevitable footdrill or square-bashing, the latter complicated by the fact that the army was changing over from "forming fours" to "forming threes". Although all recruits were normally expected to learn musketry — as rifle-drill was still officially described — a shortage of rifles made this impossible. Some rifles were supplied to Caw Camp to enable the guard to be armed and a number of gunners were mobilised as full-time soldiers to perform guard duties. One of these was Tommy McCready who even today is not certain whether the rifles would have been safe to fire! At Ballymena 26 Battery was even worse off; they did not receive rifles until September 14th — eleven days after war had been declared. However the six rifles supplied from Derry allowed 26 Battery to take over guarding its own camp from the Ulster Special Constabulary.

Few of the men who volunteered to join the Regiment had any military experience so that everything was totally new to them. They had come from all walks of life and represented as broad a cross-section of their community as it was possible to find. They joined knowing that there would be a war sooner or later and they were prepared to make a contribution, however small, to the defeat of Nazi power. Initially they undertook to join the Supplementary Reserve for a period of four years with an extension to five years in the event of a national emergency.

For this they were to be paid the princely sum of £6 per year (non-tradesmen) rising to £15 per year for Trade Groups A and B. To qualify for this pay a minimum of twenty drills per year had to be performed in addition to attending a fifteen days camp. Additional payments in the form of proficiency grants could be made while attendance at a training camp was paid at army rates of pay. The total amount that could be paid to a trained man who was not a specialist but who had attended camp, completed all of his proficiency drills, with an extra thirty added, and had passed his weapon training or equivalent test was £11; this was made up of £3 proficiency grant, 50p weapon training, £1·50 extra drills (i.e., 5p, or one shilling, per drill which lasted one hour) and £6 bounty. Two weeks in camp for a Gunner would have added only another £1·40 (fourteen shillings per week) although food would have been provided and a family allowance paid to married men. It was hardly a financially attractive proposition even in 1939!

By the summer of 1939 the Regiment was almost up to strength and training was going on apace at Caw, at Ballymena and at Coleraine. Uniforms and boots had been supplied and each man was learning how to look after his kit and to bull his boots to produce a gleam sufficient to keep a sergeant-major happy. The uniform they received was the old Service Dress rather than the newly-introduced battledress with its blouse type jacket. Caps were the field-service caps worn at an angle on the right side of the head and the badge was the bursting-grenade rather than the full cannon badge. In fact the initial issue of uniform had included only a set of overalls, a belt and a field-service cap per man!

Physical fitness was regarded as a priority and route-marches and physical training programmes were used to produce a high level of general fitness.

The training programmes were also used to identify potential NCOs. Apart from previous military experience the only other criterion available initially was to select people who had experience in supervision or management in civilian life, and a number of selections were made on this basis although those concerned received only provisional rank. As the training progressed however it became possible to identify more accurately those who could make good NCOs. Successfully completing a PT course in Belfast resulted in Jim Canning being promoted to Bombardier — the artillery equivalent of corporal. Indeed he had finished top of his class and when he later passed a drill course he moved up to Lance-Sergeant.

Not all of the old soldiers were prepared to become NCOs. One who refused sergeant's stripes and chose to remain in the ranks was J J Doherty, who had served in the Royal Irish Fusiliers and Irish Guards and whose reserve commitment had only ended in December 1938. Since he was a Class 'A' Cook he was offered the job of Cookhouse Sergeant. He had been a reluctant cook however having been remustered in India following a serious illness which left him temporarily unfit for normal infantry duties. Instead "J J" became a Despatch Rider with 25 Battery — a job which suited a man who loved motorcycles. As 25 Battery was to suffer badly from their cooks in the years ahead it was perhaps as well that J J Doherty kept his refusal to be Cook Sergeant quiet!

As that last summer of peace wore on the organisation of the Regiment was coming together extremely well. Each of the heavy batteries was sub-divided into four operational units. At first these were called "Troops", but they later became "Sections" and were designated A, B, C and D Sections, 24 Battery, E, F, G and H Sections, 25 Battery and I, J, K, and L Sections, 26 Battery. At Coleraine 6 LAA Battery was divided into X, Y and Z Troops. Each section was commanded by a junior officer who reported through the Battery Captain to the Battery Commander. In each of the heavy batteries the Battery Captains were "home-bred" — 24 Battery had Captain G S Maxwell, 25 had Captain Sir Basil McFarland, Bt, and 26 had Captain Sam Haughton. Sir Basil McFarland had also been elected to the Mayoralty of his city.

Various officers and NCOs had attended a variety of courses leading up to the Regiment's own scheduled practice camp in Scotland in September. During August a further Predictor and Height and Range Finder arrived for 24 Battery at Caw but hardly had they been received than all of the Regiment's guns and instruments were sent to 8th Regiment in Belfast for the defence of that City during the "crisis" which arose as German forces menaced Poland in the final week of August 1939. As 9th Regiment's guns and instruments went to Belfast their parent formation, 3rd AA Brigade, issued the Regiment with its first vehicles! 8th Regiment had been deployed on a war footing on August 25th.

On Friday September 1st 1939, Poland was invaded. Mobilisation orders went out to the army and its reserves. When war was declared, on Sunday September 3rd at 11.00 a m, 9th Regiment was already completing its mobilisation plans. The Regiment was now "embodied" into the Regular Army and all personnel were called out as full-time soldiers. Training and preparation intensified as the Brigade deployed for anti-aircraft tasks in Northern Ireland.

Recruiting continued for a time after mobilisation and there was a significant number of volunteers after September 3rd. These included men like Percy Hamilton, an ex-Inniskilling Fusilier, Miles Stott, and Bertie Cuthbert, the latter being only 18 years old at the time. Those who volunteered after September 1st did not qualify for

Colonel W R Shilstone, MBE, RA, the Commander of 3rd Anti-Aircraft Brigade, inspects 24 AA Battery at Caw Camp in October 1939. The young officer on the right of the picture is 2/Lt Arthur W Hogg who was to be the final Supplementary Reserve Commander of 24 Battery in 1946.
(Photo: McDonald/Bigger Collection)

Colonel Shilstone inspects 25 AA Battery at Caw.
(Photo: McDonald/Bigger Collection)

the Efficiency Medal after the war as this was granted only to TA or SR members — the former with a "Territorial" clasp, the latter with a "Militia" clasp — and both organisations had become part of the regular army upon mobilisation.

When the Regiment was mobilised only one man failed to report for duty and it was later discovered that the Irish Army had a prior claim on his services! Medical examinations declared all but one man fit for service — although at Coleraine a man with an artificial eye had passed his medical test!

During this period elements of each of the Batteries of 9th Regiment deployed to Belfast to assist 8th Regiment and to improve their own training. Here they came across the 3·7 inch gun for the first time as well as the Sperry Predictor. However, they found that at times the equipment was non-existent or failed to work. At Orangefield Second-Lieutenant Arthur Hogg of 24 Battery found himself on a gun position with no fuze-setting tool — and the position might as well therefore have been minus a gun — so he found a tool-maker who could fabricate the required instrument in case the gun had to be used in anger.

Back in Derry it had not proved possible to bring all of the men in to live in Caw Camp due to the shortage of accommodation and cooking facilities. They were therefore "dispersed to their homes", although some were accommodated in Ebrington Barracks. A programme of calling one Troop in each day for duty had also had to lapse because of the difficulties involved. Lectures and footdrill had featured on this programme as well as PT courses, and arrangements simply had to be revised to suit the prevailing conditions.

On October 20th a farewell dance for the Regiment was held in the Guildhall hosted by the Deputy Mayor. The Regiment had received its warning order to move to England for "an unknown destination" in early November. Already however rumours of a move to the Middle East were circulating! Present at the Guildhall dance were a group of members of 25 Battery who had been recalled from Belfast that day. They had been amazed to receive a reception as warm as if they "had won the war" upon their return. That Guildhall dance was a prelude to a series of functions in the City throughout the war years to provide cash for the Regiment's Comforts Fund — a solid indication of the place the Regiment already held in its native City.

On October 21st the Regiment's Honorary Colonel, His Grace the Duke of Abercorn, Governor of Northern Ireland, inspected the Regiment at Caw Camp before going on to lunch in the Officers' Mess which was in the Northern Counties Hotel. At lunch the Duke told the officers of how he had been told "confidentially" by one sergeant that "we are going to Egypt but the officers don't know yet". Whatever his source of information the sergeant was to be proved absolutely right.

Small advance parties from Regimental HQ (RHQ) and each of the heavy batteries moved across to Hampshire to Deepcut Barracks to prepare for the arrival of their comrades. Then, on Saturday November 4th 1939, RHQ, 24 and 25 Batteries marched out of Caw Camp and, accompanied by the band of the South Wales Borderers, made their way to the Midland Station where they boarded the 4.30 p m train. There were scenes of great emotion as the City bade farewell to its soldier sons

The Limavady detachment of 24 Battery pictured at Limavady prior to the Regiment's departure.
(Photo: Jack McIlmoyle)

HM Troopship *Ettrick* on which the Regiment sailed from Marseilles to Alexandria.
(Photo: Author's Collection)

and the train pulled out on its way to Larne. Joined by 26 Battery from Ballymena they crossed to Stranraer and travelled by train to Farnborough in Hampshire from where they marched to Deepcut.

The Regiment was now out of 3rd AA Brigade which was destined for France with the BEF. However they had left behind their light battery and at Deepcut they were joined by 5 LAA Battery. This unit belonged to 8th (Belfast) Regiment but it had been at practice camp in Lancashire and was ordered to move to Deepcut on November 5th to join 9th Regiment. Members of the Regiment believe that 5 LAA Battery joined them by mistake — someone is alleged to have written '5' instead of '6' on a deployment plan — but it seems likely that it was simply a matter of administrative convenience to move a Battery already in England to Deepcut. 5 LAA Battery's War Diary simply notes that the Battery was "attached unofficially to 9th Regiment, RA (SR)".

A week was spent at Deepcut before moving to West Firth Barracks, Blackdown where the Regiment passed another week. Much of the time was spent in foot-drill, lectures and route marches and kit was issued for their move.[3] As the items issued included topees (pith sun helmets) and light khaki drill (KD) uniforms the rumours of a move to join the Middle East Forces seemed to be confirmed. Innoculations were also carried out as if to reinforce the conclusion that the destination was somewhere quite hot!

On November 18th they marched from Blackdown to Farnborough North station where they entrained for Southampton and crossed the Channel in what crew members described as the worst storm for many years. They were very happy to disembark at Cherbourg early next morning and grateful that they had not joined the Navy! Most of the day they were free to sightsee in Cherbourg before once again boarding a train that would take them through the night across France.

The journey took them through Le Mans, Vierzon, Bourges, Saincaize, Moulins and Paray to Marseilles. En route they were issued with rations on the train — tea, bread and tins of herring (one tin to two men). Conditions were cramped — some men even sleeping on the overhead luggage racks to allow more room on the floor for their companions. A number of stops were very much appreciated in order to stretch their limbs. Meals were served at some of the stops and Bertie Cuthbert recalls the spectacle of what might have been soup being served out of horse-troughlike containers on the platforms. The offering had cooled by the time he received his portion which was topped by a layer of unappetising grease. His share was tipped out over the platform. At that time he did not realise how typical of army food this concoction actually was.

At Marseilles on November 21st at 9.00 a m they boarded HM Troopship *Ettrick* and sailed under escort for Alexandria. En route they put in at Malta for three hours

[3] There was also sport particularly football and Ted Griffiths recalls that as sports officer of 5 LAA Bty he arranged a match between his Battery and 26 Bty. Prominent in that game was the giant Blair Mayne who made a dash for 26 Bty's goal with the ball tucked between his chin and his chest. Needless to say he scored!

on the 24th before continuing eastwards. A programme of sports and lectures filled their time on board and on the 27th they arrived in Alexandria. After disembarking they moved to the eastern outskirts of the city to the tented encampment of Sidi Bishr where they experienced a sandstorm for the first time. Until February 1942 Alexandria would be home to the Regiment.

CHAPTER TWO

ON EGYPT'S SHORES

Sidi Bishr Camp had been established during the Great War but in 1939 it was still a tented camp with only two permanent buildings — the guardhouse and the cook-house, both of which were of wooden construction. It had until recently been occupied by 7th Medium Regiment who had moved out to allow 9th Regiment to take over.

Due to the sandstorm on the first evening there was an air of disarray and it was the following day before the gunners had a good opportunity to look around. The tents were set in sand with duckboards across the sand for paths and with boards as the floors of the tents. Cooking and eating meals was a long process and much time was spent in the early days until the difficulties of feeding almost one thousand men were overcome. Among those problems was the serious digestive ailments encountered by many of the men. One bad attack of stomach pains with the symptoms of dysentery was tracked down to the use of baking soda to clean cooking utensils without proper rinsing. Within days "steps had to be taken for the proper supervision of the cleaning of fruit and vegetables", such was the effect of improperly prepared food on the gunners.

When Christmas arrived the cooks had still not got their act under control. In spite of a grant of £70 from the Comforts Fund — made up of donations from home — the Regiment found itself without a Christmas dinner. The discontent expressed by the rank and file was such that cooks and food were brought in from the Coldstream Guards at Mustapha Barracks, thus ensuring that this first Christmas away from home should not be memorable for the absence of a festive dinner! Christmas greetings were received from the Deputy Mayor of Londonderry on behalf of the citizens.

The Regiment had come to Alexandria to provide air defence for the port, which was the Royal Navy's principal anchorage in the Eastern Mediterranean. Airfields in the area, chiefly Dekheilia to the west and Aboukir to the north-east, were also guarded by AA detachments. Interestingly the city of Alexandria was not considered a prime target for enemy aircraft. This arose out of the peculiar situation stemming from Egypt's neutrality; although the Kingdom of Egypt broke off diplomatic relations with Germany in 1939 it remained neutral until February 1945 when it declared war on Germany. British forces were present in Egypt as a result of the Anglo-Egyptian Treaty which gave the UK rights in the Canal Zone and in Alexandria. Thus British guns in Alexandria were there to defend British ships and shore establishments rather than the city.

In December 1939 the gun defence of Alexandria was provided by 20 Anti-Aircraft Battery, RA, and by detachments of Royal Marines. There were also Egyptian Army AA defences in the area. Both 20 Bty and the Marines began to hand over positions to 9th Regiment's batteries and to help train those batteries prior to the Regiment assuming the main responsibility for the ack-ack defence of the harbour. On December 2nd, 26 Battery's I and J Sections took over two positions from 20 Battery and eight days later the Regiment had a General Officer Commanding's (GOC) inspection.

Two days after Christmas both 24 and 25 Batteries moved into gun positions. 24 took over the No 1 Position at Fortress HQ, Ras-el-Tin, which included four 3·7 inch mobile Mark 1A guns, a Sperry Predictor and a Goerz height and range finder. Half of 25 Bty moved out to the north-west of Alexandria to the Agami position which was occupied by Royal Marines. The remainder of 24 and 25 Batteries continued to reside at Sidi Bishr. In addition to the No 2 Position at Ras-el-Tin Palace 26 Battery had also occupied No 7 Position at Dekheilia, which was both Alexandria's airport and an air station for the Royal Navy. The Regiment's vehicles, which they had last seen in England, finally arrived in Alexandria on December 28th.

During January 1940 the Regiment assumed responsibility for the positions they had taken over. At Agami 25 Battery replaced the Royal Marines completely and began to construct additional accommodation. Indeed by the end of the month most of the Regiment had moved out of Sidi Bishr. Regimental HQ had been sited at El Silsila on the promenade fronting the Eastern Harbour and was christened Londonderry Camp. However C and D Sections of 24 Bty together with 5 LAA Bty remained at Sidi Bishr, the latter because none of their guns had arrived. Although the Battery had recce'd positions for its Bofors 40 mm guns these did not arrive in Alexandria until May 15th 1940 — for training 5 LAA had to borrow a Bofors from the Egyptian Army.

In February 25 Battery moved into Dekheilia where Battery HQ was established. However they were forced to move out of this accommodation towards the end of March due to a large influx of Fleet Air Arm personnel. A new camp was then established at Mex close to Seagull Hotel and dubbed Seagull Camp. Major Frank Dearden had been taken ill in March and was invalided home at the beginning of April. His departure was much regretted by his men who made a presentation to him before he left. Captain Sir Basil McFarland assumed temporary command of the battery.

May saw further changes in the defences and during the month 20 AA Bty was placed under the command of 9th Regiment. 20 Bty, which had previously been independent, possessed six rather than four sections and could thus man three positions. These were at Mustapha (No 5), El Mafrouza (Quarantine Station) (No 16), and Old Lighthouse Bay (No 17). For a time they had had a position at Sidi

The Mechanical Transport (MT) Section of 25 AA Battery at Sidi Bishr Camp, December 1939. In front from L to R are Bobby McFeeters, Bertie Porter, Tommy Gamble (MT Sergeant), J J Doherty and Bobby Simpson. At back are (L) Jimmy McFeely (Toby) and Bertie Monteith (Monty).
(Photo: Author's Collection)

Alexandria from the air showing the Gare du Caire in the foreground. Just above the gardens in the bottom left corner can be seen Kom-el-Dik fort which later became a gun position.
(Photo: Author's Collection)

Bishr but the sections involved then moved to No 16. 24 Bty established a new position (No 13) at El Silsila with two guns released from Ras-el-Tin where Mk2 static guns were being emplaced. A few days later El Silsila was brought up to four guns with the arrival of another pair of static 3.7s manned by C and D Sections who finally moved out of Sidi Bishr.

At Agami 25 Bty handed over two of its guns to the Royal Malta Artillery and moved to a new four gun position (No 8) at Agami Point. This had been built by the battery and BHQ moved there also in early June. 26 Bty moved from No 2, where they were replaced by B Section of 20 Bty, on June 6th and from Dekheilia to Aboukir east of Alexandria (I and J Sections) and to Port Said and Port Fouad. E and F Sections of 25 Bty had gone west, literally, on May 7th when they had moved up to El Daba in the Western Desert.

5 LAA Bty had disposed eight of its newly arrived guns in Alexandria at the harbour, at the floating dock and at the oil storage tanks at Gabbaria. A further four guns went out to Aboukir. Of those in Alexandria itself two were actually placed on the outer breakwater so that they might be used to engage mines as they were dropped by parachute into the water.

On May 14th 1940[1] the Regiment was notified of the War Office decision to approve the title of 9th (Londonderry) Anti-Aircraft Regiment. Although this had been used unofficially it was nonetheless pleasing to have official blessing for this territorial title. Then on June 1st came another change when the Regiment officially became a Heavy Anti-Aircraft regiment rather than simply an Anti-Aircraft regiment.

There was a second 5 Bty under the Regiment's command also. This was 5 HAA Bty, Royal Malta Artillery which arrived in Alexandria on May 9th and by the end of the month occupied positions at Aboukir (No 15) and at Wardian (No 4).

Thus it was that the end of May 1940 saw the entire Regiment at a high state of readiness. At the same time the British Expeditionary Force was being evacuated from France and with it the other units of 3rd (Ulster) AA Brigade (SR). What had been dubbed the "Phoney War" by the press had ended in Europe; it was about to end in the Mediterranean too.

Since arriving in Alexandria the Regiment had devoted a large proportion of its time to training. Within two weeks of their arrival officers and NCOs had participated in the Air Defence of Egypt exercises held in December 1939. As soon as the gunners moved into positions they began to train on the guns and instruments on a daily basis.

Anti-aircraft gunners ideally needed aircraft to assist in their training. This air co-operation was provided from a number of sources some of which were not always entirely satisfactory. Civilian contracts had been placed for this task but the aircraft supplied were often unsuitable and indeed at times unavailable. The other source of aircraft for air co-operation was the Royal Navy. Naval aircraft proved to be more reliable but again they were not always suitable. Later the Naval aircraft became less

[1] The official date of this decision was May 9th 1940.

reliable as they were called upon to perform similar tasks for the Fleet which of course received priority. Arrangements could therefore be cancelled at short notice which was highly frustrating.

Neither the civilian aircraft nor the naval machines were capable of steep diving and thus could not be used to simulate diving attacks. Nor did their speed permit them to represent adequately the modern bombers which could appear over Alexandria — they were in fact some 100 mph *slower* than those bombers. Eventually in 1941 the job of air co-operation was taken over by the RAF who provided a more reliable service with faster aircraft.

The air co-operation planes were used to practise both shooting and the use of instruments. Shooting was carried out against a target or sleeve towed at a safe distance by the aircraft — one naval pilot is reputed to have signalled to the crew of a warship carrying out anti-aircraft exercises "I am towing this *xxxxxxx* thing, not pushing it!" The aircraft itself could be used as the target provided a round with a half-charge was used in which case the shell would explode well short of its mark.

Height and range-finder and predictor crews also practised against these aircraft. Again because of the lack of speed the training was not always as realistic as would have been wished but it did give some worthwhile experience to the crews. Fire control personnel were able to go aloft in Fleet Air Arm machines to check the accuracy of the height finders; for most of them, like Alex McCluggage of 26 Bty, it was their first experience of flying.

While this training programme had been going on the gunners had also been working at their positions. Most of the manual work was in fact carried out by themselves and thus they had little rest in those early days of 1940. One of the few days off had been St. Patrick's Day 1940 when a Sports Meeting had been organised and additional rations provided for the evening meal from the Comforts Fund.

Each of the gun positions now operated as a self-contained unit. At each position there were four gun-pits normally with concrete revetments strengthened by sand-bags. The four guns would be controlled by a Command Post (CP) which was linked to the guns by land-line (telephone) by way of a relay bunker manned by two telegraphists outside the CP. An officer was normally in charge of the post and was designated Command Post Officer (CPO) with perhaps a senior NCO as his assistant (ACPO). The remaining CP personnel included the spotter, the height/range-finder crew (two) and the predictor crew (six) as well as the relay telegraphists in their bunker — a total of about fourteen men. Ten men manned each gun under the command of a sergeant (or a bombardier) and Despatch Riders stood by in case of a breakdown in the communications system. Each position also had men to carry out guard duties, driving duties, cooking and all the other tasks that were part and parcel of the life of a gunner regiment.

All of the men lived at their gun positions either under canvas or in dugouts. There was no luxury for them anywhere and food and recreation were also provided at the sites with a tent doubling as both mess and recreation hall. When they heard that they were being described by the traitor Lord "Haw-Haw" as "Featherbed

Swordfish of the Fleet Air Arm over Alexandria. Some of these aircraft were shore-based at Dekhelia and were sometimes used for the calibration of heightfinders.
(Photo: Tom Moore, Ballymena)

Height 10,000. A Height and Rangefinder in use at Agami Point, 1940. Manned by 26 HAA Battery.
(Photo: Tom Moore)

Soldiers" their reaction was to grin and shrug it off. As far as nicknames were concerned they had another — in his telegrams home Sir Basil McFarland would report "all Derry Boys with me are well" and "Derry Boys" they remained. The title was accepted by the Ballymena men of 26 Bty, the Ards men of 5 LAA Bty, and by the English, Scots and Welsh who joined them — a first draft of reinforcements having already reported.[2]

In addition to having no featherbeds they were generally unable to avail of many of the luxuries still obtainable in Alexandria, even though there had been some time for sightseeing in the city. An ancient and beautiful city, Alexandria, or El Aksandr,[3] took its name from Alexander the Great and was steeped in history. For the off-duty serviceman the Co-Ordinating Committee for the Welfare of HBM Forces in Alexandria published a "Services Guide to Alexandria". This included a brief historical outline of the city, a list of services clubs in the city, details of hotels, churches, sports facilities, cinemas, public transport and the cost of postage to the UK.

At the Catholic Soldiers' Club in Mustapha Barracks "eggs, liver, pork, beef, fish and English sausages are served with chips, onions, tomatoes, bread and Australian butter and a good cup of tea made with fresh milk at prices ranging from PT 3½ to 5½ per plate". (from about 7d to 1/= or 2½p to 5p). Bed and full board was 25 PT (piastres) a day (25p) or 30 PT for Lance-Sergeants and above at the Royal Arms on the Route de la Corniche while Talbot House — TocH — provided a bed for 7 PT (7p) with breakfast, lunch and dinner at 8 PT (8p) each. Service Freemasons recognised by the Grand Lodge of England were invited to get in touch with resident masons at the Masonic Hall on the Rue Toussoun Pasha.

In today's terms the prices quoted seem incredibly cheap. To the ordinary soldier of 1939/40 this was not so. A gunner had a pay of 14 shillings (70p) per week and from this a sum of money was alloted to his family at home leaving him with about 7 shillings (35p). Further deductions had to be made for essentials such as laundry and the army charged for "barrack damages"! (Tommy McCready wonders how he ever damaged a barracks since he spent all of the war, except for two weeks, in a tent, a dug-out or a hut.)

There was not therefore much money left to spend on entertainment of any form and thus it was hardly surprising that much of their entertainment was made by themselves at their positions. Money could of course be saved for occasional leaves which could be taken in Cairo using hotels or even in Alexandria at a leave centre in Mustapha Barracks. Although there were occasional "binges" in Alexandria that city did not see as much of the "Derry Boys" hard-earned cash as Lord Haw-Haw would have had his listeners believe.

Many of the soldiers also spent some of their free time in the Sacred Heart Convent in Alexandria, where there was a community of Irish nuns of the Franciscan Order. Sister Mary Richard Coyle from Rosemount was a very young nun teaching in Alexandria at the time and she recalls the many men from the Regiment who visited

[2] This draft also included Channel Islanders who went to 5 LAA Bty.

[3] The modern spelling is Al Iskandaríyah.

the Convent — everyone was welcomed by the community and many lasting friendships struck up. Today Sister Richard, now in New Jersey, USA, recalls men like Jack Gault — "a gentleman who came to visit me at home in Derry when I returned for the first time in 1947" — Willie Carey, a neighbour from Rosemount, Matt Mulhern, from the Waterside, Bobby Mitchell from the Fountain, Tom Patton from Strabane (who had lived in Derry before the war and played football for Rosemount under the management of Dan Coyle, Sister Richard's father) and not least a quiet Tyrone man, J J Doherty, who was to become Sister Richard's brother-in-law.

While all of the training, the digging and the building had been going on the Regiment had been fully operational and on a war footing. Initially each of their positions had been at the "third degree of readiness" whereby half of the sections at a position were at four hours notice and could be allowed away on duty or on leave for a four hour period. The other men could, apart from duty NCO, guards and telephonists, be allowed to leave the position for work, games or baths, but could not go more than one mile away.

Then as the likelihood of Italy entering the war became greater readiness was upgraded to the second degree described as "the normal war routine" when all positions were expected to have sufficient men at their posts to man the equipment fully in thirty seconds. Men surplus to the requirement for the full and efficient manning of equipment could leave the position on duty or on leave for not more than two hours at a time.

The "first degree of readiness" was that adopted when a raid warning was received or an attack expected. All personnel would then "stand to" action stations ready to engage a target as soon as it appeared.

Italy declared war on the United Kingdom on June 10th 1940. Alexandria and the "Derry Boys" were now in the front line.

CHAPTER THREE

A LA CARTE — ARMY STYLE

It has virtually been a tradition in the army that soldiers grumble and there is no doubt that the favourite topic for a good grumble has been food. Soldiers love to tell the story of how a cook did more harm to a unit than the enemy, or of the men who saved an entire battalion/battery/squadron with a single shot — by shooting the cook! The cause of these grumbles and stories — army food — was never available in sufficient quantity and was, generally, the same monotonous fare.

Part of the blame, for the monotony at least, can be laid at the door of the French commissariat which, during the Napoleonic wars, experimented with a method of providing a ration-meat that was palatable for the front-line soldier without having to resort to the then normal system of supplying large numbers of cattle "on the hoof" to a field force. They did this by boiling beef and putting it into glass jars. Later the glass jars gave way to tins and when the idea crossed the Channel to Britain and tinned boiled beef became a staple part of the soldier's diet the French word "*bouilli*" (boiled) was anglicised to "bully". So bully-beef was born.

"Bully" was normally served up for the main meal of the day but there were times in some units where due to the exigencies of service it appeared for breakfast, dinner and supper! Fortunately that did not happen too often.

For Ninth's gunners a typical day would start with breakfast which often included porridge made from biscuits and perhaps a fry using tinned bacon and soya — an item of diet that proved to be as unpleasant to the taste as it was to the eye. Eggs appeared now and again but were usually on the menu for special occasions. The biscuits for the porridge incidentally were sealed in tins and preserved in malt and when the tins were opened the smell of malt could be detected a good quarter of a mile away. At Sidi Bishr in 1939 and early 1940 tins of biscuits which had been in store since 1917 were opened despite the labels which declared that they should have been used by 1927!

The main meal of the day was "world without end bully" beef served up in a variety of ways — stewed, fried, curried or in a pie and accompanied by vegetables which were more likely to be tinned than fresh. Not very much imagination was shown by the cooks, most of whom were not cooks anyway but simply gunners forced to do a job they disliked. On the advance to Tripoli in January 1943 Billy Smyth suggested to one of 25 Bty's cooks as they began to prepare the evening meal that the boys might like pancakes. "Let them eat bully"! was the reply; and bully they ate.

The bully and veg were accompanied by potatoes which were never off the menu for long. Except that is for several months in the Sudan when 25 Bty thought that potatoes had become extinct! Certainly the local substitute did not satisfy Irish palates as good wholesome spuds would have done.

Would you believe that the cooks also prepared a pudding each day? More times than enough it was rice for the "duff" as it was called by the soldiers, but now and again it might be something a little more appetising — perhaps even a pudding baked with flour.

The third meal (and meal sequences could vary according to circumstances) was "tiffin" which could be served at lunchtime or in the late afternoon. "Tiffin" was a cold meal which would comprise some peculiar mixes such as fish and pickle with perhaps cheese and biscuits — those army biscuits again! There was margarine to spread on the biscuit to help the taste but the margarine (oleomargarine to give it its full title) was a ghastly substance, especially in the heat of North Africa. James Lucas remembers that he and his infantry colleagues used to use it as oil for their trench lights!

Every meal was accompanied by mugs of tea — at least that was how the cooks described this particular fluid. Those who drank it will recall that it was called "tea" if sugar was added or "soup" if salt was added. Whichever description applied it tasted always of onions! North Africa veterans remember that, apart from "Bully", the one thing that seemed never to be in short supply was onions.

Fish often replaced "bully" in the main meal menu and of course appeared at other times as well. Tinned fish proved to be a useful item for barter with American soldiers later in Italy. The American rations included "Spam" or spicy ham which made a pleasant change from "bully". As no fish at all was included in the GIs' diet the "Yanks" were always amenable to exchanging "Spam" for fish thus making everyone happy.

Of course it was the monotonous appearance of "bully" on the menu which was the principal cause of its unpopularity. On the opposition side the Axis troops had a ration meat which was canned in Italy and proved equally unpopular. It was a sausage meat but the Germans gave it the nickname of *Alte Mann* (old man) from the initials AM which were stamped on the tins. They claimed that it was actually old men from the workhouses of Berlin and elsewhere who had been recycled into army rations! James Lucas was a POW for a time in Tunisia and had to suffer a daily diet of AM (which also stood for *Arme Mussolini*, or Poor Mussolini); he says that the most unattractive aspect of AM was the fact that it was preserved in a violet-coloured gelatine and appeared most unappetising although the meat "was quite palatable". Small wonder then that after the fall of Tobruk in June 1942 the Axis soldiers were delighted to find that their booty included huge stocks of "bully". In his book "With Rommel in the Desert" Heinz Werner Schmidt recalls how well they ate for weeks on this delicious meat and wondered how the British soldiers complained so much about it! Parcels of "bully" were sent home to Germany and Italy where it was regarded as a delicacy but, writes Schmidt, after a time feasting on "bully" they too began to grow weary of it and could see that, to the Tommies, it was just as *Alte Mann* was to the German or Italian.

Some of the "Derry Boys" may have been fussy eaters at home but they soon learned to eat whatever they got in Egypt for there never was enough to satisfy their hunger. Of course they didn't have sufficient money either to supplement the army rations except on occasions when they had saved enough to buy a meal or when they could get a cheap meal at a Forces Club. In their first week at Sidi Bishr they were actually sifting the sand between the boards of their tent floors to see if the previous occupants had lost any "akers" (money) in that sand!

But on special occasions such as Christmas Day and St Patrick's Day extra messing allowances would be made from the Comforts Fund to provide a better menu. What could be done on such days is illustrated by the 24 Bty menu for Christmas Day 1941 which is reproduced on the opposite page.

The "Derry Boys" found that they got no sympathy from their English, Scots or Welsh comrades when they did grumble about the food. These men had been drafted into 9th Regiment as the war went on and almost all of them were conscripts. Their retort to any complaint about food, or anything else for that matter, from an original member of the Regiment was "you volunteered for this". There was no answer to that, even for a Derryman!

24th H.A.A. BATTERY R.A.

No. 1 GUN POSITION

0700 hours	-	Holy Communion
0730 ,,		Breakfast
0830 ,,		Parade Service
1230 hours	-	Dinner
1700		Tea
2000		Concert Pantomime

A printed programme and menu for Christmas Day in Egypt gives a hint of all the careful preparations made for a special occasion.

MENU

BREAKFAST

PORRIDGE
BACON AND EGGS
TEA OR COFFEE
TOAST AND MARMALADE

DINNER

STUFFED ROAST GOOSE   ROAST PORK
WHITE ONION SAUCE
NEW POTATOES   GREEN VEGETABLES
XMAS PUDDING   MINCE PIES

Beer and Minerals

Fruit and Nuts

TEA

CHRISTMAS CAKE
MINCE PIES AND JELLY

SING FOR YOUR
SUPPER

At great expense we have secured the services of the well known "Stooges" who will present: —

"CINDERELLA"

A Pantomime with a Moral

14 Scenes — Wonderful Dresses — Wigs by "Scrim"

ALL SINGING TALKING DANCING

XMAS DAY 1941

CHAPTER FOUR

"ITALY IN"!

Modern Italy had become a nation as recently as 1870 following the endeavours of Count Cavour and the legendary Garibaldi. The contemporary sign of greatness among European nations was the possession of overseas territories — "a place in the sun". The new Italian state, as the inheritors of the Roman tradition, felt obliged to build an Empire but opportunities were limited. Germany, another nation which became united only in 1870, also joined in the race and the extent of Italian gain was Italian Somaliland and Eritrea in Africa. However, as the first decade of the twentieth century came to an end Italy was able to expand its African possessions. As Turkish hegemony collapsed in northern Africa Italian troops invaded Tripolitania and Cyrenaica. Italy had a foothold on the southern shore of the Mediterranean which was later ratified after the Great War. When Mussolini became dictator of Italy in 1922 he continued the programme of Italian domination across the Mediterranean by improving communications and taking punitive action against the rebellious Senussi. From the capital, Tripoli, an excellent road was built along the coast leading right up to the border with Egypt. As with the great roads of Roman times it was intended to help control the country and to assist in future expansion.

In 1935 Italy attacked Abyssinia and in a bitter, bloody campaign, condemned by the League of Nations, ousted Emperor Haile Selassie from his throne. Thus the Italian North African Empire consisted of Libya, Abyssinia, Eritrea and Italian Somaliland. With the surrender of France and the isolation of Britain Mussolini saw the opportunity to add to that empire. The glittering prize of Egypt with the Suez Canal and the Anglo-Egyptian Sudan beckoned him — Italy could soon possess all of North-east Africa. His military advisers implored him not to attack Egypt — they argued that it would be 1943 before the army and air force would be well enough equipped to knock out the British. *Il Duce* reckoned otherwise. The British, he said, were beaten; their forces in Egypt would be demoralised and easily defeated. Marshal Balbo, the Governor in Libya, was ordered to prepare an invasion of Egypt. The Italian Navy would soon have possession of Alexandria and the Mediterranean would indisputably be "*mare nostrum*" — our sea.

On June 10th 1940 Mussolini declared war. The War Diary of 26 Battery for that day notes "Italy in". Next day British patrols crossed from Egypt into Libya and RAF bombers struck at Turin. On June 28th Balbo was killed by his own anti-aircraft guns at Tobruk and Marshal Rodolfo Graziani succeeded him. It would be September 13th before Graziani could finally be forced to invade Egypt but before this Italian forces had attacked British Somaliland, Alexandria had been bombed and the guns of 9th Regiment had fired in anger for the first time.

The first Italian aircraft arrived over Alexandria on June 12th. Before dawn broke an aircraft could be heard circling overhead but it was not picked up by the searchlights and fire was not opened. In the afternoon three aircraft flew very high over the harbour but due to cloud and poor visibility they could not be engaged.

Ten days later the first bombs fell. At Ras-el-Tin, El Silsila and Agami Point 24 and 25 Batteries went into action during an alert lasting five hours from 1.00 a m to 6.00 a m. At Aboukir 26 Battery had some 18 bombs fall in its area, two of which hit the Officers' Mess. Later in the day an Italian reconnaissance aircraft was engaged by the Agami and Ras-el-Tin guns and it was afterwards reported to have crashed, being credited by the RAF and RA Headquarters to 25 Bty — the Regiment's first "kill".

July proved to be a busy month for the men on the heavy guns. 24 Battery at their positions at Ras-el-Tin and El Silsila recorded twelve enemy raids or reconnaissance missions during the month. In the Canal Zone 26 Battery was based at Port Said, El Gamil and Port Fouad (K and L Sections) with I and J Sections at Aboukir. Three raids were recorded at Aboukir, but the Canal Zone guns, although ordered to "stand-to" on occasions, did not fire against enemy aircraft and continued the work of training and improving the positions with the welcome assistance of Royal Engineers.

The month's first raid had occurred at Alexandria on July 3rd when ten Savoia Marchetti SM81s attacked the harbour. Flying in close formation at 16,000 feet they were picked up at 7.55 a m by the El Silsila position which then lost them in clouds. 24 Battery's position at Ras-el-Tin fired nine rounds at the aircraft before the clouds hid them from view there too. The Italians dropped a number of bombs, about nine of which fell in the harbour area. Later that morning an Italian reconnaissance plane was engaged by the Ras-el-Tin guns. Again nine rounds were fired and the aircraft changed course and dived to avoid the gunfire.

Four nights later both Ras-el-Tin and El Silsila positions of 24, and 25 Battery at Agami, swung into action against a recce plane just after 9.00 p m. Two of the guns at El Silsila achieved bursts on the tail of the aircraft which had been illuminated by searchlights. The stricken plane, streaming black smoke, dived for several thousand feet before it was lost by the searchlights. Its loss was later admitted by Italian radio, giving both 24 and 25 Batteries a "kill" apiece. Three of the four guns at El Silsila fired a total of twelve rounds. Ras-el-Tin's guns had also fired — three of them putting up six rounds. The Number One gun at Ras-el-Tin had been unable to engage due to its proximity to Ras-el-Tin lighthouse. In fact the entire position with its command post and four gunpits was crammed into a very confined area between the lighthouse and the water. Next morning the lighthouse again precluded the use of No 1 Gun when the Ras-el-Tin position fired on a reconnaissance plane at 8.17 a m. This happened again on July 9th at 1.15 p m when another recce machine came over, dropping some bombs for good effect.

If Ras-el-Tin's lighthouse imposed restrictions on the operations of one of the guns there was also a problem at the El Silsila position. This in fact had been condemned as unsuitable on June 11th by Brigadier A J S Pollock, the Middle East AA Commander, on the grounds that it was too open to sabotage and the activities of fifth columnists. The location of the El Silsila guns meant that the promenade had to be closed whenever the guns were about to fire — even for practice! Pollock and the Battery Commander, Major Lambert, reconnoitred an alternative position on El

Silsila Point but the guns were not moved and a further question about the suitability of 13 Position, El Silsila was raised by Brigadiers Pollock and Graham on July 23rd.

Enemy aircraft were engaged on July 13th and 15th by the guns at Ras-el-Tin; on the latter occasion bombs were dropped in the Great Pass and the Drunken Beacon.

Bombs fell again on July 16th, 24th and 26th but as the raids were during the night and the searchlights failed to pick up the enemy aircraft no guns opened fire.

Daylight raids towards the end of the month were engaged by both of 24 Battery's positions and by the Agami position but with no positive results. The same was true for I and J Sections of 26 Battery at Aboukir who fired on enemy bombers on July 4th and July 16th. 26 Battery were by the end of the month preparing to move K and L Sections from the Canal Zone to Agami Point while I and J Sections moved from Aboukir to Wardian. G and H Sections of 25 Battery left Agami for Port Said on August 6th but their stay there would be short as just over two weeks later they would move south to Port Sudan on the Red Sea to join E and F Sections which had gone there in June after their return from the desert.

25 Battery were now out of the Alexandria area completely while 5 HAA Battery, Royal Malta Artillery, under command of the Londonderry Regiment, concentrated at Aboukir at Nos 14 and 15 Positions. The other 5 Battery — 5 Light AA — had yet to fire in anger, as the enemy aircraft which had attacked Alexandria so far had remained well above the effective ceiling of their Bofors guns.

There was a total of seven raids during August and it was in the course of one of these that the Regiment suffered its first fatal casualties. At 4.20 a m on August 26th Italian bombers struck at Alexandria Harbour. 24 Battery's guns at Ras-el-Tin and El Silsila opened fire as did those of 26 Battery at Wardian. It was across this latter position that a stick of eight bombs fell, one of them exploding in the Command Post area. The blast killed two gunner/telegraphists who were positioned to link the Command Post with the gun sites. The dead were Gunners John (Jock) McMeakin and George Millar while Sergeant Harry O'Neill was very seriously injured, losing an eye as a result.

On August 31st Lt Col Smith issued the following Special Order of the Day —

"The names of Number 1449826 Gunner Millar, G
and
Number 1487270 Gunner McMeakin, J
26 AA Battery, RA, are placed at the head of the
REGIMENTAL ROLL OF HONOUR
"The Commanding Officer on behalf of all ranks, wishes
to take this opportunity of expressing his sympathy
to the relatives and comrades of these men, and to
record his appreciation of the magnificent spirit with
which the Officers and men of I and J Sections 26
AA Battery RA continued to engage the enemy after the
tragedy occurred".

Such was the spirit that the Sections showed that one man, Sergeant Norman Neilly, was later mentioned in despatches for his calm and inspiring behaviour in the aftermath of the attack. The dead gunners were buried that afternoon with full military honours in the British Military Cemetery.

At Agami K and L Sections of 26 Battery were in action for the first time on August 15th when the Battery Captain, Sam Haughton, was pleased to note that they were "reasonably steady under initial fire".

The combined fire of 24 and 26 Batteries totalled 124 rounds during the month and there were again a number of occasions on which the 3·7s could not engage. One of these occurred on August 15th when for the first time enemy aircraft attacked at low level. At Ras-el-Tin an attacking aircraft was illuminated twice by searchlights but was out of range of the guns on the first occasion and on the second was below the safety angle of the guns at the lighthouse. At Wardian I and J Sections of 26 Battery fired one round from a 3·7 but also brought their light machine guns into use, hosing 2,304 rounds of ·303 ammunition from Bren guns against two Italian bombers.

At long last 5 LAA Battery was able to go into action and fired 27 rounds from four different positions. No hits were claimed by any of the defenders and some bombs fell in the harbour. Four days later 5 LAA Battery were given the task of training two batteries from an anti-aircraft regiment of the Australian Imperial Force (AIF) on Bofors guns.

On September 13th the long-awaited Italian attack on Egypt was finally launched. Graziani's army moved across the frontier and occupied Sollum. The force of five infantry divisions supported by about 120 tanks and first class artillery attacked through the Halfaya Pass, later to be renamed "Hellfire" Pass by British soldiers, took Sollum, which had been occupied by a platoon of British Infantry, and pressed on eastwards. However, the advance ground to a halt at Sidi Barrani only fifty miles further on. There they began to dig in and, unbelievably, went on the defensive. A total of three corps, twelve divisions strong, occupied a fifty mile coastal strip of Egypt. British troops in Egypt were greatly outnumbered and yet the Italians failed to press home this advantage. One very good reason was that the Italian Tenth Army was not very mobile, lacking sufficient mechanised transport for its infantry, and it had not been trained in desert warfare, unlike its opponents. Apart from their artillery their *matériel* was not good and was out-classed by the British equipment. However their navy was first class and might have been used to support the land forces. It wasn't, nor was the air force — the *Regia Aeronautica* — used to best advantage.

In Alexandria the defences expected a heavy onslaught during September. This did not materialise and the entire month saw only five raids on the harbour. 24 Battery fired only 22 rounds during September in raids on two days early in the month (September 6th and 8th), 26 Battery fired 29 rounds, most of them on September 6th, and 5 LAA Battery did not fire at all during the month.

There were once again occasions when because of the failure of the searchlights to pick up aircraft no engagements occurred. One of these happened in the early hours of September 21st when two delayed action bombs were dropped in the Eastern Harbour near the El Silsila position. One of them exploded 4 days 22 hours and 25 minutes later causing no damage.

Even though there had been some fairly determined attacks by the *Regia Aeronautica* no damage had yet been done to ships in the harbour. This had not gone unnoticed by the naval authorities and indeed the Resident Admiral at Alexandria had already attended a practice shoot by 24 Battery at El Silsila. During this time also urgent reinforcements for Wavell's Middle East Command had been sent through the Mediterranean to Alexandria on the direct orders of Winston Churchill. Wavell now had some 150 new tanks, 48 additional anti-tank guns, 20 more Bofors light AA guns and almost 50 more field-guns. Churchill had also countermanded Admiralty plans to pull the Mediterranean Fleet out of Alexandria completely and had indeed ordered its reinforcement. The role of 9th Regiment in defending Alexandria Harbour would continue to be vital.

General Wavell still expected a further Italian advance and was preparing to meet this at Mersa Matruh. It soon became obvious however that Graziani had no intention of moving forward. In spite of his overwhelming advantage in numbers he had been shaken by the British resistance to his advance during which his casualty figures were some ten times greater than those of the Western Desert Force which opposed him. As he had been one of the voices advising Mussolini against going to war his nerve was now considerably shaken. To shake him even more Wavell planned a limited counter-attack to be launched before stronger Italian or German forces could reinforce Graziani. This attack would not happen until December and in the meantime units of Western Desert Force continued to harass the Italians in their positions in Egypt.

The remaining months of 1940 saw sporadic attacks on Alexandria. In October 24 Battery fired only seven shells at enemy aircraft during the month. All of those were fired on the night of the 16th when a number of enemy aircraft raided the harbour and several were illuminated by searchlights; one used its machine guns to fire at the searchlights. Other occasions saw the Battery unable to engage because the searchlights could not pick up the attackers — on October 21st over twenty explosions were counted at El Silsila as bombs fell close to the position in the Eastern Harbour. Next day the searchlights illuminated an aircraft but it was out of range when it was picked up. Three days later this episode was repeated.

26 Battery were much busier. Early in the month I and J Sections moved to Agami and K and L Sections came down to Wardian. The latter position fired 54 rounds during the month and the Agami guns put up 40. October 16th was a particularly hectic night at Agami with three separate engagements and bombs falling close to the positions. No casualties were sustained nor was any damage caused to the Battery's guns or equipment. Five aircraft were engaged by Agami on October 22nd, one of which was probably hit; that morning Wardian had fired 17 rounds at an enemy bomber.

An RAF bomber was fired on by 26 Battery on October 27th but fortunately it was correctly identified as a Blenheim after only one round had been fired.

During October there had also been a visit to 24 Battery's position at El Silsila by a Turkish Military and Naval Mission accompanied by Rear Admiral Elliot, Brigadier Pollock and Colonel Azim Bey of the Egyptian Army. The mission also visited Regimental HQ at Londonderry Camp, El Silsila. Later in the month the Secretary of State for War, Mr Anthony Eden visited the Middle East on a fact-finding tour for the Prime Minister. He paid a visit to the Agami position on October 24th where he inspected the guns and spoke to all present; some members of 24 Battery were also included.

5 LAA Battery had also been in action during October. Again this was during the frantic night of October 16th when four of the Bofors to the south and south-east of the harbour fired 103 rounds.

November was to provide more "trade" for 5 LAA Battery with guns in action on three occasions — 17th (13 rounds), 19th (37 rounds) and 24th (42 rounds). There was a definite upturn in enemy activity during the month shown by the fact that 24 and 26 Batteries together fired some 1,000 rounds, almost two-thirds of that total coming from 24's guns.

In mid-October a fifth section had been formed within 24 Battery and it was ready to relieve 'D' Section at El Silsila before the end of the month. Its presence at El Silsila meant no reduction in strength when C Section with two 3·7 inch guns moved to the Western Desert on November 6th where the Section's role was to help defend the Western Desert Force as it built up for the forthcoming offensive.

At Matruh C Section joined a section of 16 HAA Bty to provide a 4-gun position and experienced a number of raids during which they came to the conclusion that the searchlights at Matruh were "as bad if not worse" than those at Alexandria! In both locations the searchlight crews were mainly Egyptian with a British detachment. There were two very heavy raids at Matruh but no claims were made for aircraft destroyed or damaged. One of the raids saw eight bombs fall near one of the searchlights and a bombardier from 8th Field Regiment was killed.

The result of another raid was to improve the quality of the Section's food. Robert Hamilton recalls how tightly rationed the food was until the morning after a raid when Tommy McCready saw hundreds of fish floating dead near the shore — killed by blast. Needless to say they ate plenty of fish for several days. C Section returned to Alexandria in early December after a month in the desert.

In addition to its own fifth section 24 Battery was reinforced by L Section of 153 HAA Battery, 51st HAA Regiment, with two 3·7 inch mobile guns which were based at El Silsila from December 9th to 27th turning No 13 into a 6-gun position. The reinforcement was hardly necessary as only one raid occurred in December. In the early morning of December 19th a number of aircraft were engaged by barrage fire from El Silsila and Wardian, a total of 26 rounds being fired. Once again bombs fell near the Wardian position but although damage was caused no injuries occurred.

From the beginning of November Regimental HQ had introduced a barrage scheme for the defence of the Western Harbour. This reduced the guns' reliance on the searchlights — something which was rapidly becoming a sore point. Although there were some who felt that the barrage scheme was wasteful of ammunition it would prove its worth time and time again in the future. The Alexandria barrage involved putting up a semi-circular pattern of shells at given heights and thus there are comments on nights such as November 16th that there was no work for the predictor and height-finder teams.

Radar had also been introduced since July 1940 with GL sets (Gun Laying radar) coming on to the sites. It did not always work but as the war progressed the equipment became more sophisticated and more reliable.

By the end of December Graziani's army had been driven from Egypt. Wavell had launched Operation COMPASS on December 8th. The build-up had been disguised as a training exercise and what was originally intended as a "raid" caught the Italians reeling at Sidi Barrani. The first British offensive of the war took Sidi Barrani in hours with 20,000 Italian prisoners. Lt General Richard O'Connor, commanding Western Desert Force, decided to go on despite Wavell's withdrawal of one of his divisions[1] and moved into Libya, taking 38,000 prisoners and capturing 73 tanks, 1,000 vehicles and 237 guns.

Included in that advance were the guns of a troop of 5 LAA Battery which had been withdrawn from Aboukir on December 5th to join Western Desert Force and by the end of the month they were located at Sidi Barrani (2 guns) and Buq Buq (2 guns).

Christmas 1940 was celebrated by the Regiment in places as far apart as Sidi Barrani, Alexandria, Port Sudan and Atbara on the border between Sudan and Abyssinia. In the year about to finish they had gone through their baptism of fire, had suffered their first casualties and inflicted their first damage on the enemy. The pattern of army life in war had also been established with periods of relative inactivity interspersed with episodes of hectic action. Their skills as gunners were being refined and even when they were not engaging the enemy they practised and practised and practised on guns and instruments until serving the guns became second nature. Despite the pressure of enemy attacks on Alexandria (and at Port Sudan) no ship under the protection of the Regiment's guns had yet been damaged! It was a good start for any regiment.

CHAPTER FIVE

THE ACK-ACK DEFENCE OF ALEXANDRIA

At the time of Italy's declaration of war the total of Britain's AA artillery resources in the Middle East consisted of eight obsolescent 3 inch guns, thirty-two 3·7 inch guns and twelve 40 mm Bofors guns. Of this total of fifty-two weapons, 9th (Londonderry) HAA Regiment provided all of the 3·7s (as 20 HAA Bty had come under the Regiment's command) and Bofors guns, thus indicating the level of responsibility which had fallen upon the Regiment.

Until then the War Office had been directly responsible for overseas AA units but this system, remote enough in peacetime, was obviously not going to work in a

[1] O'Connor's troops were depleted because Churchill had ordered Wavell to send a force to Greece, an intervention which the Greek government was wary of lest it bring the Germans into the campaign.

war situation. Recognising this Whitehall appointed a Brigadier, Anti-Aircraft (BAA) for the Middle East and on May 18th 1940 the first appointee to the post arrived in Alexandria. All AA and coast defence (CD) resources in the Middle East theatre were placed under the command of Brigadier (later Major-General) A J S Pollock — all that is except the guns of the Egyptian Army (although an Egyptian liaison officer was appointed to Pollock's staff to assist in the co-ordination of the Egyptian guns with those of the Royal Artillery), those of the South African army and any AA guns deployed under the command of a subordinate formation eg an Infantry or Armoured Division.

Brigadier Pollock's task was an unenviable one as he was faced with increasing demands for AA protection. He simply had not the resources to satisfy these. The Royal Navy wanted guns to defend the Fleet at Alexandria, the army wanted them to defend their lines-of-communication to and in the Western Desert, while the RAF felt that more protection should be provided for their airfields. Alexandria's claims were strong enough to merit a Regiment — the 9th being stationed there (although within the "Alex" area the Regiment was also protecting airfields) — leaving the balance of Pollock's scanty inventory of guns to be spread among the other claimants.

The calls on the AA guns were increasing almost daily as the navy asked for guns in the Canal Zone and Sudan pleaded for weapons to be sent there. Thus 25 HAA Bty found itself to be probably the most widely-dispersed battery in the entire Royal Regiment of Artillery in the early summer of 1940 (see Part 2; Chap. 1). Fortunately however the situation was beginning to improve. 5 HAA Bty, Royal Malta Artillery arrived in Alexandria to bolster the defences and were added to 9th Regiment's strength. The Maltese remained until the needs of their own island resulted in their recall there. For a time in May 1940 9th Regiment's strength included not only its own batteries, ie 24, 25, 26 and 5 LAA Btys, but also 20 Bty and 5 HAA Bty, RMA. Although 25 Bty had four guns up the desert 20 Bty had two additional sections (ie four guns) to compensate and the arrival of the Maltese considerably strengthened the defences.

The arrival of 51st HAA Regiment further eased Pollock's problems but there was still an acute shortage of fire control equipment — predictors, height/range-finders and TIs (telescopes, identification). Until July 1940 there was not even one set of GL (gun-laying) radar!

When enemy aircraft began to come over Alexandria there were occasions when the gunners wondered if they could ever have any real effect. Training had concentrated on the need for accuracy of fire, but when aircraft began to dodge about the sky such accuracy was not easily attained. Indeed picking-up the attackers was not always easy. During daylight hours a combination of height, glare and the colour of the sky made aircraft-spotting very difficult while at night-time reliance on searchlights made the situation even worse. Originally many of the searchlights were manned by the Egyptian Army whose standards of professionalism left much to be desired. John Taylor recalls the night that an Egyptian searchlight crew picked up a bomber over Ras-el-Tin — their immediate reaction was to abandon the light and begin an impromptu celebration dance! Needless to say the aircraft escaped; the

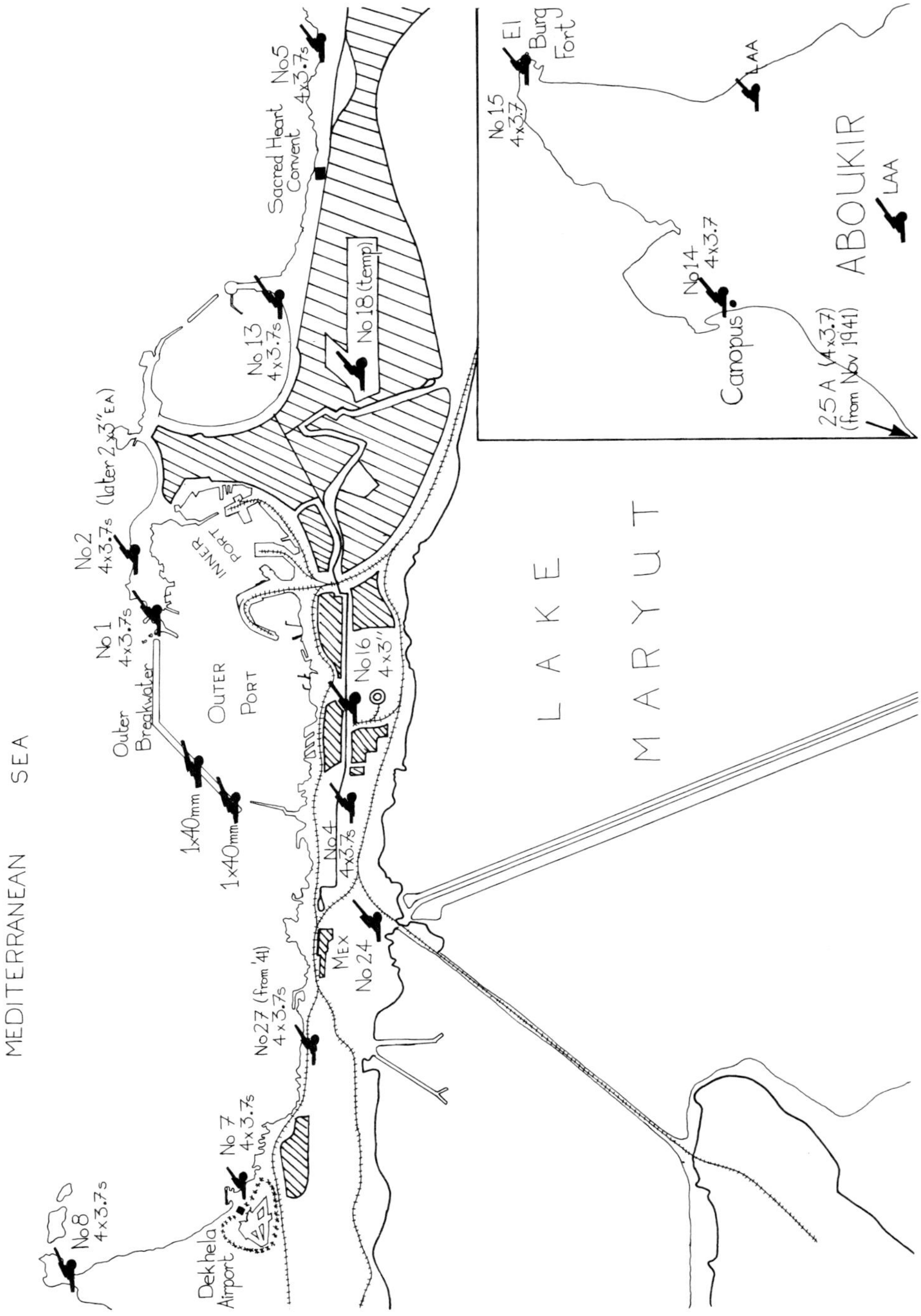
MEDITERRANEAN SEA
LAKE MARYUT
ABOUKIR
OUTER PORT
INNER PORT
Outer Breakwater
No 1 4×3.7s
No 2 4×3.7s (later 2×3" EA)
No 4 4×3.7s
No 5 4×3.7s
No 7 4×3.7s
No 8 4×3.7s
No 13 4×3.7s
No 14 4×3.7
No 15 4×3.7
No 16 4×3"
No 18 (temp)
No 24
No 27 (from '41) 4×3.7s
1×40mm
1×40mm
Sacred Heart Convent
MEX
Dekhela Airport
El Burg Fort
Canopus
LAA
LAA
25A (4×3.7) (from Nov 1941)

ALEXANDRIA GUN DEFENCES

Showing the main locations which were manned by guns of 9th (Londonderry) HAA Regiment from December 1939 to March 1942.

No 1	Ras-el-Tin:– 24 HAA Battery
No 2	Ras-el-Tin:– 26 HAA Battery, later replaced by Egyptian Artillery with two 3 inch guns
No 5	Mustapha:– 20 HAA Battery (under command of 9th Regiment from May to September 1940)
No 4	Wardian:– 5 HAA Battery, Royal Malta Artillery (under command 9th Regiment) then 26 HAA Battery
No 7	Dekhelia:– 26 HAA Battery, later 25 HAA Battery until April 1940
No 8	Agami Point:– 25 HAA Battery, later 26 HAA Battery (from August 1940 until March 1942)
No 13	El Silsila:– 24 HAA Battery from May 1940 until March 1942
No 14	Canopus:– 26 HAA Battery, later 5 HAA Battery, RMA. 25 HAA Battery took over this position in October 1941
No 15	El Burg:– 5 HAA Battery, RMA. 25 HAA Battery also took over this position along with No 14. Both remained under 9th Regiment control from 1940 to 1942
No 16	El Mafrouza (Quarantine Stn) 20 HAA Battery
No 18	Kom-el-Dik Fort:– 20 HAA Battery
No 24	Mex:– Egyptian Artillery. This was the site which suffered a direct hit on the night of June 7th/8th 1941 killing 21 soldiers
No 25A	Aboukir:– Manned by a composite Section under 25 HAA Battery
No 27	Manned by a composite Section under 26 HAA Battery (from December 1941 when No 25A was handed over to 187 HAA Battery under command of 9th Regiment)
Note:–	Nos 14 and 15 Positions were manned by 276 HAA Battery of 68th HAA Regiment. 276 Bty came under 9th Regiment command

The LAA sites marked were among those manned by 5 LAA Bty during its time in Alexandria.

Egyptian crew was replaced next day. In fairness it must be recorded that some of the early British crews were not all that hot either and there are many exasperated reports in the War Diary of 24 Bty noting that aircraft could not be engaged because the searchlights had failed to illuminate any targets. However the British Searchlight Detachment, consisting of RA gunners and men of the Royal Wiltshire Yeomanry in the main, began to improve and helped their Egyptian counterparts to do likewise.

Thus it was that the defenders of "Alex" had to fall back on the barrage system to provide an effective defence of the harbour. Such a system had been used in rudimentary fashion in the First World War and had been revived in France in 1940 and then over Britain itself. As the ammunition stockpile situation began to improve in the later part of 1940 a geographical barrage was devised for the Alexandria area.

In principle the barrage system required putting up a line of bursting shells in the path of attacking aircraft in order to force those attackers to change height or direction or even, as a bonus, to shoot some of them down. To provide a barrage defence for Alexandria it was necessary to draw up a mass of gunnery data for a series of lines surrounding the "Vulnerable Area". The decision on which precise line or lines to use in any attack was made by the Gun Operations Room (GOR) on the basis of early warning received from the RAF.

Radar helped to refine the system, although it would be much later in the war before such control became really accurate. Even so a GL Predicted Barrage was introduced as another option for the "Alex" defenders in 1941. The object of this was "to concentrate the fire of all guns which can be brought to bear on as short a barrage line as possible by employing one or more GL sets in the Gun Zone . . .". For this barrage which, with certain exceptions, was to "take priority over all other methods" the line was divided into twenty-three sectors each of 1200 yards. The instructions for the GL Predicted Barrage run into five closely-typed foolscap sheets indicating the mammoth amount of work involved in devising such schemes. Other refinements later included a "swinging, following barrage" which called upon each gun to swing through an arc at 15 second intervals and thus follow attacking aircraft; this proved to be "very effective against persistent attackers". Although the early GL radar sets were rudimentary they did prove effective — the set at El Silsila especially so — and were a tremendous advance for AA gunners.

German and Italian aircrew used different tactics, as well as different aircraft. The German or Italian machines could be differentiated by their engine sounds as well as by appearance, thus allowing those on the ground to know what to expect from the raiders. The *Regia Aeronautica* flew high, straight and level and indeed their first attack was almost a display of precision flying. Such tactics required a high level of courage, and that the Italian airmen possessed in abundance. (Incidentally both RAF Bomber Command and the USAAF bombers used a similar approach.) *Luftwaffe* crews, on the other hand, used a variety of tactics and were the first to employ low-level bombing against Alexandria. The Germans would attack with a mixture of steep or shallow dive-bombing, low fast passes and high level approaches on the same raid — all aimed to cause the greatest possible problems and confusion for the defenders.

By mid-1941 however any enemy aircraft over "Alex" could be sure of a hot reception. In fact such was the effect of the barrages put up by Ninth's batteries that by that summer intelligence sources were reporting that few attacking aircraft were able to return without either having dead or wounded on board or having sustained damage to the airframe of the plane itself. The Regiment was informed that "the enemy airmen do not like the barrage at 'Alex'".

The confidence and morale of the "Derry Boys" had increased by leaps and bounds as they saw that their efforts were having an effect and that bombers were being shot down, damaged or forced to turn away. That improved morale was shown in a number of ways — messages painted on 3·7 rounds being one example. "Derry's gift to Mussolini" was to be seen on rounds at El Silsila where Tommy McCready had also painted the slogan "Erin go bra" on one of the guns. It was a far cry from the uncertain days of June 1940 when one gunner even decided that there was no point in going on duty as "we're losing anyway"! (He rapidly had some commonsense knocked into him by his mates and duly reported for duty.)

No doubt too the fact that the Regiment was Irish also helped as Irish units have always displayed an *élan* which possibly stems from the fact that they have had to show that they are not just as good but much better than other units, especially English!

On that note it must be recorded that the least competent gunner in all of the Royal Regiment of Artillery served in 24 Bty — and he was an Englishman. He had been posted to the Battery from RA Base Depot during 1941 and 24 soon found out why he had been posted out of his previous unit. Gunner X could do nothing right; he could not be trusted to load a gun or to carry ammunition to it, he couldn't be used on any of the fire-control instruments or to guard a site — even the cooks didn't want him, such was his unsurpassed level of incompetence. After being tried at a number of tasks, and trying the patience of a lot of men, he was finally placed under supervision to answer the telephone in a command post. At last there was a job that he couldn't mess up — or could he? The day arrived when he answered a 'phone call, acknowledged it and began to log the message. The officer-in-charge enquired "What was that about, Gunner X?" "Three aircraft approaching from the north-east sir" was the reply "and I'm logging it". A few moments later he spoke again, "Must be a new kind of plane, these. I've never heard of them before". A puzzled officer asked "What kind is that, X?" "HOSTILES, SIR," was the reply!!!

CHAPTER SIX

UNDER SIEGE IN EGYPT

1941 began on a bad note for Italy. Mussolini's armies were suffering reverses in North Africa[1] and in the Balkans. Bardia in Cyrenaica fell to the Australians on January 3rd, the same day on which Greek forces, which had already pushed the invading Italians out of Greece, launched an offensive against Italian forces in Albania. Mussolini was beginning to rue failing to heed the advice of his military chiefs that Italy was not ready for a major war!

Not surprisingly the pressure on Alexandria eased considerably at this time. The squadrons of the *Regia Aeronautica* were now primarily concerned with the defence of the Italian field army which by the end of January had lost almost 100,000 men to a British force one tenth of its own strength. During the entire month only one enemy aircraft, a Savoia-Marchetti SM 79, put in an appearance over Alexandria, being engaged by 26 Battery's guns at Agami Point.

February was also quiet although there was a spate of alerts and a single Italian aircraft was shot down by an RAF fighter some 40 miles north of Alexandria on February 4th. Because of the relative inactivity of the enemy during the month the opportunity was taken to intensify training and to improve the gun positions. 26 Battery bade farewell to its Battery Commander, Major Traill, who had commanded them since inception. Traill moved to Regimental HQ to become second-in-command of the Regiment while the Battery Captain, Sam Haughton, was promoted to take command of 26 Battery. The two "Alex"-based batteries also had visits from foreign armies — 24 receiving a Greek AA officer at Ras-el-Tin and 26 a group of Polish gunner officers at Agami Point.

However Adolf Hitler had decided to come to the aid of his hapless ally both in the Balkans and in North Africa. By mid-February Western Desert Force had captured all of Cyrenaica and had destroyed the Italian Tenth Army. For the loss of 550 dead or missing and 1,373 wounded O'Connor's force had inflicted 20,000 casualties, dead or wounded, on the Italians and had netted 130,000 prisoners including seven generals. They had advanced some 500 miles and now O'Connor looked forward to moving on to Tripoli. It was not to be, for Wavell had been ordered to move troops to Greece to face the expected German attack there and thus an advance on Tripoli was precluded. O'Connor's force was stripped of troops and equipment and ordered to hold the situation in Cyrenaica.

This body-blow to O'Connor's hopes, and to those of his troops, was delivered by Wavell to O'Connor's chief-of-staff on February 12th. Ironically on that very same day, a German major-general flew into Castel Benito airfield near Tripoli to prepare to take command of a small German force of one light division and one *panzer* (armoured) division which was to move to North Africa to support the Italians. This new force

[1] Operation "COMPASS".

Man the guns! A crew of 5 HAA Battery, Royal Malta Artillery dash to their gun at the Agami practice Camp, May 1940. The Battery had just arrived in Egypt and was under the command of 9th Regiment.

(Photo: Imperial War Museum)

No 1 Position at Ras-el-Tin photographed from the top of the lighthouse and showing how cramped this position was. It was sandwiched between the lighthouse and the harbour.

(Photo: Arthur Hogg)

had been authorised on February 6th and was given the name "*DEUTSCHES AFRIKA KORPS*". Its commander was described in a British intelligence summary as "an obscure general, Rommel"!!! Major-General Erwin Johannes Eugen Rommel[2] would soon be far from obscure. The Desert Fox had arrived.

Alexandria had its first sight of a German aircraft on March 20th. This machine was identified by the El Silsila position of 24 Battery as a Junkers Ju 88K and 17 rounds were fired at it. At Ras-el-Tin it was identified as a Heinkel He 111Mk V and 25 rounds were fired. 26 Battery's two positions at Wardian and Agami fired 23 and 21 rounds respectively at what both identified as a Heinkel. Regimental HQ at Londonderry Camp also identified it as a Junkers 88. As there were some similarities between the two types, and it was the first time that a *Luftwaffe* machine had been seen in the area, the confusion is understandable, especially as the plane was flying at about 16,000 feet. A number of other reconnaissance flights were made over Alexandria by Heinkel He 111s on subsequent days however, making it more than likely that the first German engaged was actually a Heinkel. The four recce flights by Heinkels were the only activity over the city during March and this had allowed further improvements in the positions occupied by 26 Battery as well as a high level of training by both Batteries. In the latter role 26 Battery included a route march in Field Service Marching Order (FSMO) on March 12th!

There had also been time to celebrate St Patrick's Day when Regimental HQ had been pleased to receive St Patrick's Day greetings from 5 HAA Battery, Royal Malta Artillery, which was still under the Regiment's command at Aboukir. A sports meeting had been held and extra rations provided by the Comforts Fund.

When April began Rommel had already gone on to the offensive — in spite of orders to the contrary — and 5th Light Division had pushed back the depleted British forces to Mersa Brega. Supported by the Italian Ariete Division, the Germans recaptured Benghazi on April 4th, Derna on the 7th and were at Bardia on April 12th. Tobruk was now isolated and held by an Australian force, reinforced by British and Indian troops.

Next day, Easter Sunday, 25 Battery arrived back in Alexandria after their relief in the Sudan. They were due to move into a rest camp before taking over positions in Alexandria so their frustration can be appreciated when they were given two hours leave and ordered to move up the desert to Mersa Matruh that same day. For the powers-that-be it was a simpler task to detail a unit that was already on the move to keep moving than to uproot another battery and deploy it "up the blue" as the desert was known. A few days before, on April 10th, the Regiment's CO, Lt-Col Smith, had been posted to the British Military Mission to the Egyptian Army. He was replaced on April 21st by Lt-Col L C Aitken who would command the Regiment for over three years.

Alexandria remained relatively quiet in terms of aerial activity with four reconnaissance aircraft fired on. One of these, an Italian Cant Z 1007 which was engaged

[2] Rommel should already have been well-known from the Blitzkrieg of 1940 when he commanded an armoured division in France.

by all positions of 24 and 26 on April 11th was later shot down by a Hurricane and some of the wreckage was washed up near the Silsila position on Easter Monday.

Later in the month Regimental HQ evacuated Londonderry Camp and moved to a new location on Rue de la Corniche to allow the camp to be used for the reception of Royal Artillery personnel evacuated from Greece. Some of these included men of both 24 and 26 Batteries who had volunteered to transfer to 20 HAA Battery which had moved to Greece to defend RAF airfields with eight obsolete 3 inch guns. The first evacuees, 660 officers and men, arrived on April 29th and were despatched to Sidi Gabr station to move to the Royal Artillery Base Depot.

One of the men who went to Greece with 20 HAA Bty was Tom Patton. A Strabane man, Patton had been working in Derry when he joined 24 Bty only to find that all of the other Strabane men were assigned to 25 Bty. A total of 34 men from 24 Bty and 32 from 26 Bty volunteered to join 20 Bty for its move to Greece, which was carried out by warships. The Battery travelled on the cruiser *HMS BERWICK* in a squadron of five ships, led by *HMAS SYDNEY*, arriving at Piraeus in early November 1940.

When the withdrawal was ordered the 20 Bty men fought in a rearguard role with New Zealand troops from Argos down to Monemvasia on the southern tip of Greece from where they were evacuated to Crete. Once again retreat and evacuation was called for and Tom Patton was one of the gunners who, with New Zealanders, got away from Crete on *HMS AJAX*, of River Plate fame. "After surviving a few hectic attacks from the *Luftwaffe* we arrived safely in 'Alex' early in May 1941 to be billeted in Londonderry Camp where we were warmly welcomed and entertained by our former comrades", he recalls.

George Kernoghan of 26 Bty went to Greece with his brother Willie. The Kernoghans and Alfie McClean, also of 26 Bty, were among the lucky ones who got back to Egypt. Many of the pre-war regulars of 20 Bty had been captured as were Joe Breslin, Alfie Coyle and Johnny Higgins of 24 Bty. In July 1941 20 Bty went to Cyprus where, almost as soon as they arrived in Famagusta, they were caught in an air raid by Italian bombers; several men were killed including Michael McCartney, one of the "Derry Boys" of 24 Bty. A group of ten men from 5 LAA Bty also went to Greece in the ranks of 155 LAA Bty; they did not arrive in Greece until mid-March but they were actively engaged and two of 5 LAA Bty's original members, Gunners Hugh Gamble and Willie Graham, were killed in action.

May saw a change of tempo in Alexandria. With Rommel on the Halfaya Pass and XIII Corps, as Western Desert Force had now become, holding the way into Egypt against him, Churchill ordered a special convoy — codenamed "Tiger" — through the Mediterranean to Alexandria with fresh supplies of guns, tanks, aircraft and ammunition. The convoy arrived in "Alex" on May 12th and the attention of enemy recce aircraft and bombers intensified.

In preparation for the arrival of the "Tiger" convoy *Luftwaffe* and *Regia Aeronautica* minelaying aircraft had visited "Alex" on the evening of May 1st. A number of mines were seen descending by parachute, some being fired on by the Bofors guns of 5 LAA Battery. About seven aircraft were engaged in this raid. 24 Battery's guns fired 446 rounds, those of 26 Battery 150 rounds and 5 LAA Battery put up 152 rounds. Until the convoy arrived on May 12th there were a number of recce flights and the guns of all three batteries were in action on the 4th with the heavy batteries firing on two other occasions. On the day of "Tiger's" arrival, there were two engagements eleven minutes apart. Both were made by 24 Battery at El Silsila, the first being a radar guided engagement and the second a visual in which Ras-el-Tin also participated.

Surprisingly the harbour was left alone for four days until at 12.46 p m on May 16th a Cant Z 1007 was engaged at between 25,400 feet and 26,200 feet by both 24 and 26 Batteries. At El Silsila a total of 76 rounds were fired but after firing 12 rounds the No 2 gun suffered a serious accident when the next shell exploded in the breech killing Gunner J J (Jack) McLean instantly. The blast blew Tom Moore out of the gunpit and when he had gathered himself it was to find that three others had been seriously injured, including Willy Hunter, the gun commander, who was in hospital for some days afterwards with knee injuries, while other members of the crew had to be treated for shock. John Taylor was one of the men who volunteered to clear the gunpit after Jack McLean's remains had been removed. It was a most unpleasant experience he recalls and they wondered indeed how it was that more of the crew had not been killed. In fact only the charge for the round had exploded — 8½ lbs of cordite — and the shell itself had to be discharged by ordnance personnel whose job no-one envied!

The El Silsila position was in action again two days later firing on an SM79 recce plane just before noon. There were a further six reconnaissance sorties before the end of the month and a bombing and mine-laying raid on May 30th. During this latter attack Ras-el-Tin's guns fired 112 rounds and El Silsila's 31 in barrages on the orders of the Gun Operations Room (GOR), although no aircraft were illuminated by searchlights. Bombs were dropped on Dekheilia and on the harbour boom and mines fell in the Great Pass. 26 Battery's guns also engaged with 119 rounds from Wardian and 105 from Agami Point. Once again these were predicted barrages ordered by GOR.

Next night, just before midnight, 26 Battery engaged an enemy aircraft on GOR orders. Then the searchlights caught the intruder as fire continued. Wardian had fired 56 rounds, Agami had fired 22 and when the aircraft was next seen it was reported by 14 Position at Aboukir (5 HAA Battery, Royal Malta Artillery) to be on fire and losing height. This was 26 Bty's first "kill".

Since the earliest engagements the effectiveness of the guns had increased. No longer did they remain silent if the searchlights could not illuminate a target. Now the positions were firing predicted barrages and on information supplied by gun-laying (GL) radar. The barrage system had been first introduced towards the end of 1940 and was being progressively refined. Its basic concept was to create an area of intense fire which would force attacking aircraft to break away. To protect Alexandria harbour the guns would put up a barrage based on a fixed line around the floating dock at heights up to 20,000 ft. and calculated on an attacking speed of 250 mph by enemy aircraft. Barrages could be fired at a height and in an area agreed with the

RAF thus allowing fighter aircraft to operate in concert with the guns.

There were however limitations on the operations of the heavy guns around the harbour. They were not permitted to engage aircraft over the harbour below 2,000 feet with high-explosive shells or below 1,000 feet with shrapnel rounds. Similar restrictions did not apply to ships of the fleet in "Alex" harbour! When they opened fire with their two pounder "pom-poms" the rounds fell to earth if they did not hit a target. Invariably they fell at El Silsila and on Londonderry Camp presenting what the gunners felt was an "unnecessary hazard". Protests to the navy fell on deaf ears however and they continued to blaze away with the "pom-poms" at every opportunity!

Radar had also played its part in the fine-tuning of the AA defences. In addition to long-distance radar for early-warning purposes each gun position had an anti-aircraft radar known as a GL set. These had been introduced from mid-1940, and although there were teething troubles they soon began to show their worth. Some of the problems persisted and required the presence of an Instructor — Fire Control to check that the sets were being used correctly. Problems had occurred with equipment at Nos 4 (Wardian), 8 (Agami Point) and 13 (El Silsila) Positions where the sets were at times confused. Part of the solution lay in clearing the ground around the GL set to reduce or eliminate ground "clutter". The area around the set at Agami was cleared completely within a radius of 50 yards and obstructions which protruded above 10° were removed within a radius of 90 yards. The El Silsila GL set was placed on the end of El Silsila Point free from ground "clutter" but that did not eliminate the other problems such as a reduction in the signal strength under certain light conditions — natural or artificial — or when fabric-covered aircraft were being tracked. Solutions to those problems lay in the progressive development of radar itself rather than in "on the spot" modifications. Nonetheless the El Silsila set proved to have a remarkable range being capable of detecting approaching aircraft at ranges of up to 60,000 yards. Enemy aircraft at heights from 12,000 to 20,000 feet could be picked up at 50,000 yards and false alarms could be averted if approaching "friendlies" had their IFF (Identification — Friend or Foe) transmitters switched on.

This improvement in the defences was to be put severely to the test, for the month of June was to see attacks of unprecedented ferocity on Alexandria. Rommel was again on the move, having stopped Wavell's counter attack ("Operation Brevity") at Halfaya Pass. By the beginning of June Rommel had retaken the Pass and was poised to strike into Egypt. Alexandria, the home of the Mediterranean Fleet, (and a key supplies port for the British Middle East Forces) was now a priority target. Thus it was that June saw a total of fourteen bombing or reconnaissance raids on Alexandria. Both 24 and 26 Batteries were in action during these raids. 5 LAA Battery engaged the raiders on six occasions and 5 HAA Battery, Royal Malta Artillery was also heavily involved until its relief later in the month by 276 HAA Battery of 68th HAA Regiment.

The heaviest raids occurred on June 4th and during the night of June 7th/8th. On June 4th at least nine hostile aircraft on a high-level raid dropped high-explosive and incendiary bombs using flares to illuminate their targets. Unfortunately their aiming was badly off for many of the bombs fell on civilian areas. Homes were destroyed and fires broke out, and before the all-clear had sounded 102 civilians had been killed. The guns had engaged on predicted barrage and on GL control. Five high-explosive bombs dropped near 26 Battery's position at Wardian which nonetheless fired 96 rounds; Agami's guns fired 19 rounds while those of 24 Battery at Ras-el-Tin fired 106 rounds. El Silsila position fired 63 rounds and six of 5 LAA Battery's 40 mm guns in the harbour area fired 114 rounds. The effects of this raid included a marked display of hostility by Egyptian civilians towards British service personnel, on whose presence they blamed the deaths. In fact it appears that the enemy bombers thought they were hitting the harbour area as they had been fired on on their approach by light guns. The planes moved southwards believing these to be the guns in the harbour area when in fact they were Egyptian Army guns south of the city. Thus the bombs fell in the residential areas.

At 10.56 p m on June 7th the next big raid began. It lasted, with a break of one hour, until 5.06 on June 8th. During that time a succession of bombers, operating singly or in pairs, dive-bombed, bombed from high-level or carried out minelaying attacks in the harbour area. One Heinkel He 111 was illuminated by searchlights and engaged by 24 Battery's guns but with no definite results. Yet again the bombers hit civilian targets and over 400 people died. There were military deaths too — near Mex Pumping Station No 24 Position was manned by Egyptian Army personnel and one of their sites was hit, killing 21 soldiers and wounding 19 others. 9th Regiment's guns fired a total of 750 heavy rounds (582 from 24 Battery and 168 from 26) and 180 light rounds.

Next afternoon an Italian reconnaissance aircraft, a Cant Z 1007, appeared over Alexandria. Despite its height of 28,400 feet 24 Battery opened fire and scored a hit. Jack McIlmoyle recalls the elation of the Ras-el-Tin gunners as the Cant started to trail smoke. Arthur Hogg, who commanded one of the sections at Ras-el-Tin, confirms that the position's guns scored this hit. As the aircraft lost height it was fired at by one gun of 5 LAA Battery (who claimed to hit it though this was unlikely in view of its height) and then by an RAF fighter which administered the *coup de grace*. Two crew members were seen to bale out near Agami Point but one parachute failed to open. The evidence of 5 LAA Battery's War Diary confirms that the aircraft had already been damaged by 24 Battery's fire and that it would have crashed anyway even without the intervention of an aircraft. Robert Hamilton remembers that the pilot of the Tomahawk fighter which shot the Cant down visited 24 Bty the next day to tell the gunners that he would not have been able to shoot down the Italian if the guns had not already damaged it. The RAF however had no hesitation in claiming it as an unshared "kill" for 250 Squadron's Pilot Officer Hamlyn.

On June 21st a raid lasting almost three hours caused more civilian fatalities. 24 Battery were again in the thick of the action putting up over 300 rounds from the Ras-el-Tin and El Silsila positions. At Wardian 26 Battery fired 192 rounds and at Agami 84 rounds were fired. 24 Battery's light machine guns were also used, firing 168 rounds of ·303 ammunition. An unexploded bomb landed in soft sand about 10 yards from the Ras-el-Tin position. The use of two guns was suspended but, recalls Jack McIlmoyle, when a recce aircraft came over just after noon that day there was not too much concern about the bomb and all four guns opened fire.

A two-hour raid on June 23rd caused some more damage to civilian property but no deaths were recorded. Aircraft were engaged by GL controlled barrages and one, illuminated by searchlights, under control from the heightfinders. 24 Battery, 26 Battery and 5 LAA Battery fired almost 400 heavy rounds and 83 light rounds.

On June 28th and 29th there were two further raids involving high-level bombing and mine-laying. The intruders were engaged by GL barrages and also by predicted barrages. During the raid on June 28th a large explosion was heard out to sea and it was believed that it may have been an aircraft crashing after damage from barrage fire. The final day of June was quiet, almost as if the opposing forces were trying to recover from their efforts.

Wavell was now about to depart from the Middle East, his place being taken by an Irishman with an Indian Army background — General Sir Claude Auchinleck. The "Auk's" appointment to this theatre was the direct result of the failure of Wavell's counter-attack in mid-June — Operation "Battleaxe". Although initially "Battleaxe" had seen Commonwealth forces advance on Sollum, Rommel had cleverly drawn XIII Corps' armour on to his anti-tank artillery which included 88 mm guns (originally designed for the anti-aircraft role). Superior German armoured technique saw 50% of the British tanks out of action within three days, and Wavell was forced to report "the failure of 'Battleaxe'". XIII Corps had lost 122 dead, 588 wounded and 259 missing but over 100 tanks had been lost. These were part of the precious cargo despatched at great risk to Alexandria in May with the "Tiger" convoy. No matter that the tanks had not been properly prepared for desert warfare; no matter that Wavell had been forced into launching an offensive before he would have chosen to — there had been no gain and it had been another reverse for British arms. Churchill had hoped that the risks he had taken would bring better results and decided to move Wavell to India and bring Auchinleck from India to the Middle East.

Auchinleck was granted a breather as soon as he took command on July 1st. That was due primarily to Hitler's refusal to strengthen Rommel's forces, although the latter felt that he could, with extra men and equipment, advance to the Suez Canal. Hitler's decision not to reinforce the Afrika Korps was due to his planned launch of Operation "Barbarossa" — the invasion of Russia — on June 21st.

The gunners in Alexandria felt the respite too. During the entire month of July they experienced only four raids although in each case they lasted some two hours on average. A total of 1,617 heavy rounds were fired — 905 by 24 Battery and 712 by 26 Battery. Ironically 11 of the rounds fired by 24 Battery had been at a friendly aircraft — albeit a Dornier — of the Yugoslav Air Force. In addition 277 Battery under the command of the Ninth fired 515 rounds, this unit having replaced the Maltese gunners.

5 LAA Battery had also been in action twice, firing 350 rounds. Their second engagement was in the early hours of July 26th and later that day the Battery ceased to be part of 9th Regiment and became part of a newly created Light Anti-Aircraft Regiment — 52nd LAA Regiment.

However the respite which Alexandria had been afforded did not extend into August. Axis bombers were back with a ferocity equalling that displayed in June. In all, eleven raids were launched against the harbour and were met by barrages totalling 3,655 rounds from 24 and 26 Batteries.

August's busiest night was that of the 7th and 8th. At 10.19 p m a heavy bombing raid began on the harbour. Mines and high-explosive bombs were dropped in the harbour area and the guns of 24 Battery at Ras-el-Tin and El Silsila were heavily engaged together with those of 26 Battery at Wardian. 24 Battery adopted GL barrages to fire at the intruders while 26 Battery used Predicted Barrages. One bomber was illuminated by the searchlights and 24 Battery's guns fired on it. The bomber was hit and crashed in nearby Lake Maryut. John Neilly of 26 Bty recalls that the bodies of the crew had been savagely mutilated by Egyptian looters who reached the plane before the army did.[3] Another aircraft was engaged by Bren guns, and a further three aircraft were destroyed before the raid ended just after 1.00 a m on August 8th. As night-fighters were also operating it was believed that some of the doomed aircraft could have been shared by aircraft and ground fire. 24 Battery fired 354 rounds from Ras-el-Tin and 200 from Silsila while 26 Bty fired 342 from Wardian. The Agami position fired only 71 rounds.[4]

There were other intensive engagements during the month. Indeed on August 1st there had been two raids during which the combined fire of 24 and 26 Batteries totalled 574 rounds. During a raid on August 20th the Royal Navy laid a smoke screen on the harbour while 346 rounds were fired by 9th Regiment guns.

24 Battery claimed a "possible" two nights later when the El Silsila guns engaged an aircraft on GL information. As one bomber crashed this was probably 24's victim. 482 rounds were fired by the two batteries on that occasion and four nights later 692 rounds were loosed off at bombers which dropped both mines and bombs.

On August 1st there had been damage to houses and to barrage balloons, but no fatalities. One aircraft was brought down by night-fighters which had begun operating during July but which were described by the gunners as only hampering "our style without doing any damage to enemy aircraft". In fact the night-fighters, and one pilot alone, claimed to have shot down three of the five bombers brought down on the night of August 7th/8th. AA Headquarters were however equally certain that 9th Regiment's guns had claimed three of the enemy. So too, needless to say, were the "Derry Boys"!

Nine civilian casualties, three of them fatal, were noted on August 22nd. On August 31st twelve bombs fell in the city and at Wardian when one doughty He111 pilot penetrated the barrage. Although there were civilian casualties no-one was killed.

As the Germans drove deeper into Russia North Africa became almost a forgotten theatre as far as Hitler was concerned. Auchinleck was thus able to build up

[3] A similar incident occurred at Tripoli in 1943 when the body of a *Luftwaffe* airman was washed up on the shore. Before it could be removed for burial fingers were hacked off and gold teeth removed.

[4] 26 HAA Bty adopted a policy of rotating the sections from Agami to Wardian and vice-versa as the former position was not as busy as the latter.

A mine explodes in Alexandria Harbour, 1940.

(Photo: Arthur Hogg)

his forces and in September a second corps — XXX Corps — was created, and the force consisting of XIII Corps and XXX Corps became an army — the immortal Eighth Army. Lt General Sir Alan Cunningham, an Irishman, was appointed as its first Commander. Cunningham had led the forces which destroyed the Italian East African Empire and restored Haile Selassie to his throne. (The Emperor had returned from exile in England in a Sunderland flying-boat of the RAF and had been welcomed at Alexandria by a guard of honour provided by 24 Battery.[5] The "Lion of Judah" had been amazed to learn from Gunner Tommy McCready, who was one of that guard, that men had come from Ireland to fight for his throne!)

Now the pressure on Alexandria relaxed. September saw only seven raids or reconnaissance sorties, during which a total of 1,879 rounds were fired, most of these (1,123) coming from the guns of 24 Battery. Of the month's total expenditure of ammunition 1,433 rounds were fired in the early morning of September 6th when low-level and shallow-dive bombing raids were carried out on the harbour. Later in the month 26 Battery carried out mobile training while a number of practice shoots were carried out by 24 and 26. At El Silsila on September 29th a two-hour practice shoot was held with the towed target. The promenade was duly closed and the Sections of 24 Battery from both Ras-el-Tin and El Silsila took turns to fire. 'B' Section was last to fire and, to the delight of the gunners, shot the target sleeve off its cable with their second round. Next morning 'B' Section were sent to Agami Point to complete their practice shoot at No 8 position!

There were only two raids during October — both at the end of the month. During the first, on October 28th, the Regiment's guns fired 138 rounds and during the second, on October 29th, 561 rounds. No damage was caused on either occasion.

For 9th Regiment however October was notable because of the return of 25 Battery to the fold. Having served in Sudan and in the desert since May 1940, 25 Battery finally rejoined the Regiment on October 8th and were located at Aboukir. Major Thompson, who had taken over command of 25 Battery just before the move to Port Sudan, relinquished that position on October 11th and responsibility devolved on Captain Sir Basil McFarland, Bt, who would command this Battery almost until the end of the war. Sir Basil was promoted to Major during November.

By this time also the designations of the sections had changed and each Battery had sections referred to as A, B, C and D. Thus 25 Battery's A and D Sections took over No 15 position at Aboukir while B and C Sections moved into No 14 position. Battery HQ was located at Malta Camp, a reminder of 5 HAA Battery's sojourn in the Aboukir area.

Axis aircraft paid no attention to Alexandria during November and the Regiment turned its attention to training. Readiness levels of course never reduced but 26 Battery's men found themselves route-marching again, while a scheme for standardising procedures throughout the Regiment led to a round of visits by groups from each Battery to the positions of their sister Batteries.

[5] 24 HAA Bty also provided a more sombre honour guard for the funeral of Brigadier "Jock" Campbell VC who was killed in a road accident after receiving his VC in Cairo. Campbell was buried in Alexandria and, as he was a Gunner, 24 Bty claimed the right to provide the honour guard.

Nearing competion: the new gun position (No 13) at El Silsila. In the foreground is a Mk 2 static gun but in the background, immediately under the barrel of this gun can be seen a mobile Mk1 gun.
(Photo: John Taylor)

Close defence Bren gun position at El Silsila.
(Photo: John Taylor)

24 and 26 Batteries engaged single reconnaissance aircraft on December 17th and 19th but 25 at Aboukir were not disturbed. The Aboukir defences were strengthened however with a third heavy gun position (No 25A) which was manned by a composite section drawn from all three batteries. That position was taken over by 187 HAA Battery just before Christmas and the composite section moved to a new position (No 27) between Mex and Dekheilia where they came under the command of 26 Bty. Although the Regiment now had all of its own batteries under command it also had operational responsibility for 187 HAA Battery which was located from December 20th at 25A Position, Aboukir, 26 Position, Alexandria and had a section alongside the composite section at 27 Position.

As they celebrated the Christmas of 1941 the Regiment's days in Alexandria were drawing to a close. Up the "blue" Cunningham's Eighth Army had attacked Rommel in Operation "Crusader" on November 19th. Heavy fighting saw Tobruk relieved, a counter-attack by Rommel repulsed, and Eighth Army force back the enemy in Cyrenaica. Major General Neil Ritchie took over Eighth Army in the desert, the German attack in Russia froze, literally, to a standstill, and Japan attacked Pearl Harbour and Malaya. On December 24th Benghazi was captured by Eighth Army; next day Hong Kong surrendered to the Japanese.

On December 19th in Alexandria harbour the Italians "sank" the battle-ships *HMS Queen Elizabeth* and *Valiant*. Arthur Hogg recalls that a navy liberty boat ferrying sailors across the harbour that morning saw figures on a buoy. Believing them to be sailors who had fallen in the night before the boat crew effected a rescue only to discover that the "victims" were Italian frogmen. The Italians refused to disclose where they had attached their explosives and some time later both battle-ships were holed by underwater explosions. The harbour was so shallow however that the mighty ships settled evenly on their keels on the harbour bed and there was still sufficient freeboard, combined with the lack of tides in the Mediterranean, to fool the Italian recce plane that came over that day into believing the ships had not been damaged. Indeed a reception was held for King Farouk of Egypt on board one of the vessels and neither he nor his entourage were ever aware that the ship rested on the sea-bed! Both were soon repaired and put to sea again. At the end of another year 9th Regiment were still able to point to the fact that no ship in the harbour had been sunk or seriously damaged by air attack however and a Royal Navy Officer had paid a fine tribute to the Regiment's work when he told a Derry Officer "we don't worry as long as your anti-aircraft guns are firing".

CHAPTER SEVEN

MORE IRISH THAN THE IRISH

Although the War Office had granted the official use of the name Londonderry in the title of 9th Regiment in May 1940 there was no regimental depot in the city to receive recruits, train and post them as new gunners to the Regiment, as would have been the case with an infantry regiment. As part of the mammoth Royal Regiment of Artillery, 9th (Londonderry) Regiment received drafts of reinforcements and replacement personnel from the Royal Artillery's own central training organisation via the theatre base depot. Of course this meant that the new men were more likely to be English, Scottish or Welsh than Irish and, in truth, this prospect did not appeal to the "Derry Boys" at first. However, faced with a situation which they could nothing to change they resolved to do the next best thing — they would "convert" the newcomers into Irishmen! This they succeeded in doing to a very large extent and the story of one English gunner typifies what happened.

Albert Missen was a 21-year-old who found himself posted to 25 Battery in Port Sudan in 1940. Missen, who soon acquired the nickname Dan by which he has been known ever since, wondered at first what fearful crime he had committed to deserve the fate which had been thrust upon him. "After a 12 000 mile journey around the Cape, followed by 1 000 miles down (sic) the Nile in the company of an Arabic speaking Irishman called Abdul (McLaughlin) from somewhere called Ulster (sic)" he found himself in the midst of a bunch of Irishmen who appeared to regard him as being every bit as much the enemy as if a German or Italian had reported to them.

Very soon Missen discovered that his education had been inadequate for he had to admit that his schooling had not included learning anything at all about Irish history! He was left in no doubt at all as to his ignorance. Worse still he had been educated in what was obviously a cultural backwater for he knew no Irish music or songs. When it came to keeping body and soul together he hadn't a clue about buttermilk or poundies or even poteen. Worst of all was the language problem for the Derry Boys pointed out that he couldn't even speak English properly for simple words like "wee" and "the Boroo" and "wains" left him shaking his head in mystification and looking for the services of an interpreter[1].

He need not have worried. With typical Irish generosity his new companions set to with a will to cure all of these problems. It might be a slight exaggeration to say that an induction course was held, but that effectively was what happened. Pretty soon Dan Missen and the other "foreigners" of his draft were much better educated. They

[1] Poundies — a potato dish consisting of creamed potatoes mixed with onion. Served in a mound from which a hole is scooped and milk added. Also known as "champ".
Wee = small.
Boroo = the dole.
Wains = children (esp. young children).
Poteen = a spirit distilled, illegally, from potatoes.
Buttermilk = slightly soured sweet milk. A very palatable drink.

knew about poteen and its effects, they had, in Missen's words, buttermilk coming out of their ears; they knew about the Pope and King Billy and Patrick Sarsfield and Bob Williamson (a weaver to trade); they knew about Derry's Walls and just how far it was to the bridge at Drumahoe; they could recite the street names of the city as if they had been brought up on them and they learned about drowning their shamrock on St Patrick's Day and singing "The Sash" on the "Twelfth".

In short Missen and his fellows came to be part of this Irish regiment and not simply "with an Irish mob" as the first letters home to Blighty had stated. Just how much this was so was illustrated by Missen's refusal to go on leave to Cairo from 25 Battery in the Western Desert in case he might be posted elsewhere. Indeed "after four years in your company I had begun to think of myself as being as Irish as you and when that black day came in September 1944 and we had to part company it was as though I had lost my right arm. During my years in exile I have never forgotten you and have always been proud to say that I served with an Irish regiment".

Dan Missen was not unique but was typical of what happened to the men drafted into the Regiment. During their spell at home in the autumn of 1944 Sir Basil McFarland spoke of how the English, Welsh and Scottish draftees had become Irishmen. He was not exaggerating — the *élan* which marked the Ninth had woven its web of Irish magic to captivate all those who had come into contact with the Regiment. It was a magic that has never weakened; the happy memories of a regiment that was without peer have been lifelong for many men throughout Great Britain!

CHAPTER EIGHT

ARTS REVIEW

To say that war is mostly boredom or inaction interspersed with periods of action is to repeat something that has been said so often that it is virtually a *cliché*, yet it is nonetheless true. An important task therefore for any commander is to ensure that those periods of inaction do not become spells of boredom, for a bored soldier is quite simply not a good soldier. Training, both practical and theoretical, can, and must, fill part of the time, as can general maintenance tasks — "fatigues" to the ordinary soldier — but that still leaves part of the soldier's time which is not devoted either to something positive or to eating or sleeping (which most soldiers would also assert are positive activities!)

In the past the soldier was left to his own devices to fill this time and this led to a high level of drunkenness and other crimes giving the profession of soldiering a thoroughly bad name in the eyes of the public except of course, as Kipling put it, "when the guns begin to shoot".

By the time of the Second World War things had changed for the better. More official effort was made for sport and recreation and temperance was encouraged (the latter quality had in fact been officially recognised for a long time before this).

This race card shows how things could be done "in style" for the "Derry boys".

The
RAZZLE TURF and
PONY CLUB
Colours–Orange & Green

Autumn Handicap
Friday, 7th Nov.

Side Shows - - Super Attractions

RAS-EL-TIN RACE TRACK

— TOTALISATOR ON THE COURSE —

HODGES HUNT CUP. FIRST RACE 2 p.m.

1. GLAMOUR BOY	*by* Micky	*out of* Onegun
2. LATE NIGHTS	*by* Paddy	*out of* Cookhouse
3. SYMPATHY	*by* Darkie	*out of* Pity
4. MISS FORTUNE	*by* Solly	*out of* Raw Deal
5. RECONNAISSANCE	*by* Thompson	*out of* Ramleh
6. SCALDY	*by* Burns	*out of* Dock

HARBOUR HANDICAP. SECOND RACE 2-30 p.m.

1. HAPPY NIGHTS	*by* Gunner	*out of* Numbertwo
2. STELLA LASS	*by* Rotten Beer	*out of* Onions
3. TOMMYOUSTON	*by* Tough Lad	*out of* Pee Tee
4. GIBBIT	*by* Continual Cry	*out of* Gyppo
5. BROWNED OFF	*by* Two Years	*out of* Derry
6. SHUFTI	*by* Dirty Picture	*out of* Pocket

Chief Stable Lad - Mustapha Joe

BIG BILL IRWIN - Licensed Bookmaker

"ACK-ACK" PRESENTATION PLATE. THIRD RACE 3 p.m.

1. QUICK MAIL	*by* Some Hope	*out of* Army
2. HEAVEN SUPPLIED	*by* Bint	*out of* Paradise
3. JOLLY ROGER	*by* Submarine	*out of* Alex
4. ETERNAL TALK	*by* Ess Grey	*out of* Hismouth
5. TEETOTAL	*by* All Beer	*out of* Sergeant's Mess
6. EXHIBITION	*by* Naked Lady	*out of* Sister Street

DERRY VICTORY STAKES FOURTH RACE 3-30 p.m.

1. IRISH MELODIES	*by* Joeand Willy	*out of* Flutes
2. HILL BILLY	*by* Dutton	*out of* Squeeze Box
3. NUMBER NINE	*by* Dr. Limont	*out of* Sick Bay
4. HELLUVA DIN	*by* Baldy	*out of* Little Drummer Boy
5. STREAK LIGHTNING	*by* Kommandpost Korner	*out of* Jackson
6. SLOW MOTION	*by* Wog	*out of* Working

— Silver Collection for the Dean Orphanage —

In Part Two, Chapter Five we see how important sport was considered to be even in wartime, but of course not all non-active time could be devoted to sport. Like anyone else the soldier enjoys a good film or a stage revue and attempts were made to provide facilities for these.

Sidi Bishr Camp had a cinema — a somewhat erratic cinema as Tom Moore of 26 Bty noted, with anything up to six breakdowns in an hour! There were also concert parties in the days before Italy declared war which were organised by the wives of the units which had been stationed in Egypt before war broke out. Of course that ended when the families were evacuated to Palestine in the summer of 1940.

Almost everyone has heard of ENSA — The Entertainments National Service Association — which provided professional entertainment for the troops. Few of the members of 9th Regiment who have contributed to this record have any recollection of ENSA concerts however and only one, The Barnstormers, is noted in 24 Bty's records.

So what did the "Derry Boys" do for entertainment? They improvised of course, and as there was quite a wealth of talent in the Regiment's ranks they did produce some fairly good revues. In the Alexandria area both 24 and 26 Btys had their own concert parties and when 25 finally returned in late 1941 they too produced a concert party revue entitled rather appropriately "Nomads of the Nile". There was naturally an absolute dearth of women for female roles in these shows and the answer was to fall back on the tradition of the pantomime dame! During one 25 Bty concert an alert was sounded and the concert party had to dash to their gun positions resulting in the unlikely sight of a large lady lifting her long skirts to reveal army boots as "she" sprinted to "her" gun. Sgt Bill Loughlin was probably the only No 1 (gun commander) in the Royal Artillery to go into action in "drag"!

The talent possessed by the Regiment resulted in a party including 24 Bty's Sgt Matt Mulhern travelling down to Cairo on March 16th 1941 to make a recording for a St Patrick's Day broadcast. Of course the radio provided another diversion which could also raise morale, as on the occasion when the sombre BBC tones of a news-announcer proclaimed that the RAF had just shot Hermann Goering down over the English Channel. Subsequently the same announcer declared that the RAF had later discovered that they had hit not Goering but a runaway barrage balloon!

On occasions such as Christmas when the longing for home was strongest special efforts were made in the way of entertainment — though no-one would class 26 Bty's route-march on Christmas Day 1940 as entertainment!

The gunners even managed to have horse racing! — or rather they had the pleasures of betting without the horses themselves. Games could be organised even at a gun position and some sharpmindedness was shown in drawing up the cards for such entertainments. A typical 24 Bty Race Meeting Card is reproduced on page 67.

PART TWO

THE NOMADS OF THE NILE

CHAPTER ONE

25 BATTERY — THE EAST AFRICAN EXPERIENCE

On May 7th 1940 E Section, 25 Battery moved out of the Alexandria area for El Daba in the Western Desert, where they were joined by F Section and an Advanced HQ. At El Daba they came under the command of 7th Armoured Division — the famed "Desert Rats" — and were responsible for the defence of the airstrip at El Daba, LG 104, where 211 Squadron RAF were based with their Bristol Blenheim Mk1 bombers. During E and F's time up the "blue" the Battery had lost its original commander Major Dearden who had been posted home on medical grounds. In his place came Major J W Thompson[1] who was to command 25 for almost 1½ years.

The four guns of E and F Sections were the first heavy anti-aircraft protection at El Daba and thus the gunners had the task of preparing gun sites. Wood and sandbag revetments were constructed with the aid of native labour. When one of their officers told them he was organising a "Shakespearean cellar party" the gunners wondered what was in prospect for them. They quickly found out for Lieutenant John Kershaw soon had them digging a pit to store ammunition! When Italy entered the war on June 10th 1940 the native labourers at El Daba disappeared into the desert expecting an Italian attack but although there were alerts the gunners saw no action.

Four days after the Italian declaration of war E and F Sections were ordered to move to Port Sudan. On June 16th they arrived back at Sidi Bishr to prepare for their departure which was scheduled for June 18th. G and H Sections were at the Agami Point position when their fellow-gunners left Alexandria yet again.

The first part of E and F Sections' journey was by train, and then on June 19th and 20th they transferred to a Nile steamer with their guns alongside on barges. Travelling up the Nile was an adventure in itself. On several occasions the flat-bottomed steamer stuck on sandbanks. The rescue procedure was to throw an anchor forward from the bow and then use the sternwheel to crawl the entire vessel over the obstruction! Once, while they were stuck on a sandbank, an alert was sounded when an aircraft was spotted overhead at 10,000 ft but it passed on without incident.

The two sections boarded a train again on June 26th to begin more than two days travelling across the Nubian Desert. On this journey an issue of iced water was made at intervals to combat the dry arid heat, which surpassed anything most of them had met before. For one man, at least, it was nothing new and when J J Doherty stood up in the train on the second day declaring that it was time to "take my pullover off,

[1] Major Thompson had been with 20 HAA Bty which was now part of 9th Regiment.

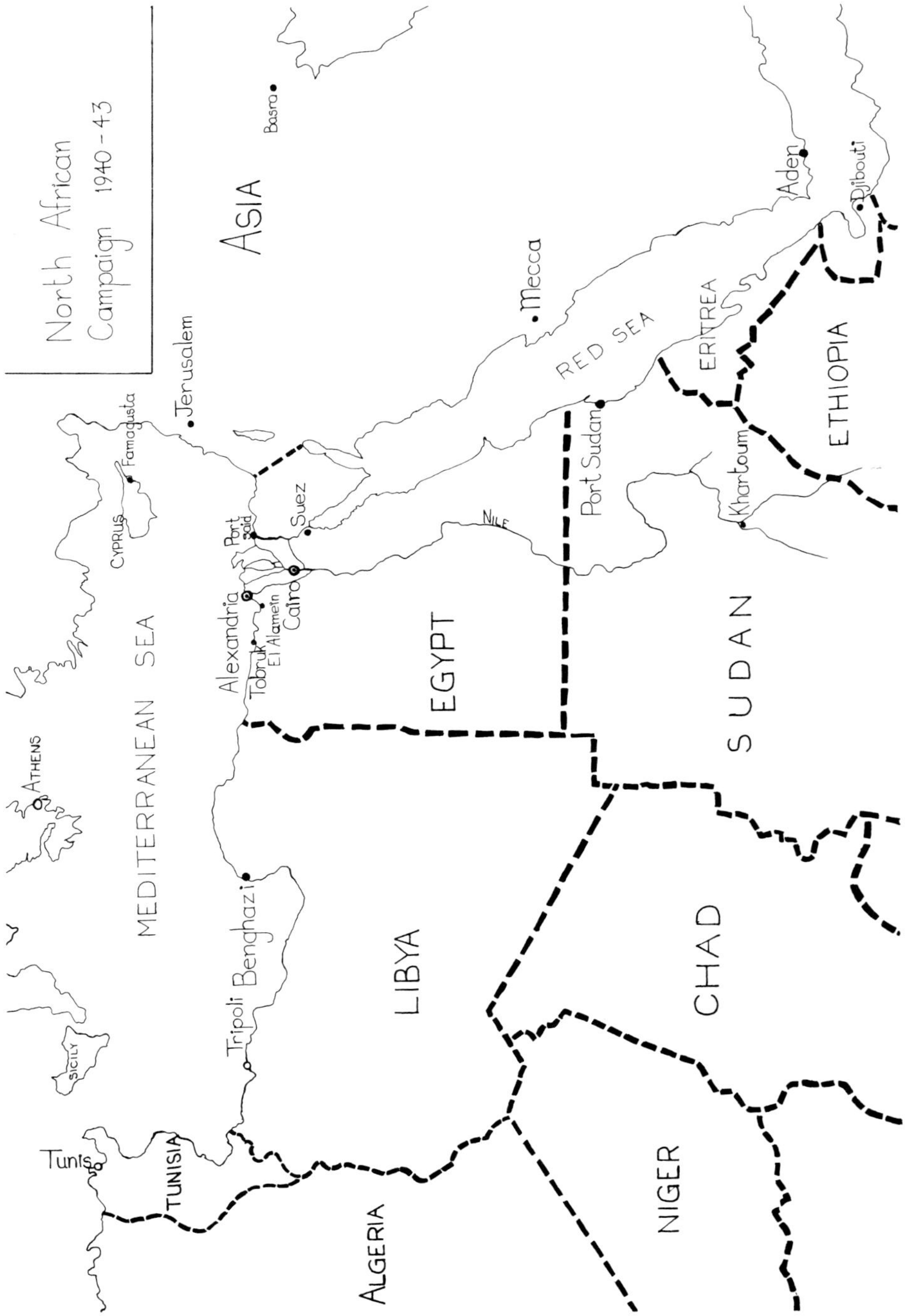
North African
Campaign 1940-43
ASIA
Basra
Jerusalem
Famagusta
CYPRUS
Mecca
RED SEA
ERITREA
Aden
Djibouti
ETHIOPIA
Port Said
Suez
Port Sudan
Khartoum
NILE
Alexandria
Cairo
El Alamein
Tobruk
EGYPT
SUDAN
MEDITERRANEAN SEA
ATHENS
Benghazi
Tripoli
LIBYA
CHAD
SICILY
Tunis
TUNISIA
ALGERIA
NIGER

boys" he was, recalls E H Babington, lucky not to be thrown out of the train. But it was all in good humour and J J (who had served in Sudan before with the Royal Irish Fusiliers) was to prove a pillar of good advice and assistance to his young compatriots.

At 4.30 a m on June 29th they arrived at Port Sudan station, and ten minutes later saw why they were needed in the area when an Italian Savoia-Marchetti SM 81 attacked the port, dropping four bombs. Although the bomber was shot down by an RAF fighter this was unusual, for the fighter defence consisted of obsolescent Gloster Gladiator biplanes which had to be already in the air, preferably with a height advantage, to have a reasonable chance against the Italian bombers.

Almost as soon as they had got themselves and their equipment off the train they began to occupy their positions. E Section was located to the east of the town with F Section to the south-west and they declared themselves ready for action at 4.30 p m, defending the harbour and the RAF airfield.

Next morning at 7.28 a m they were in action for the first time against a flight of three enemy aircraft which bombed the town. Two false alarms followed that day but, as if the Italians had been put off by the appearance of heavy anti-aircraft guns, it was over a week before another bomber appeared. In fact that machine, on July 12th, flew off before the guns could engage.

The respite had given the gunners the opportunity to build up their positions, an operation which created its own false alarm on at least one occasion when blasting operations[2] at F Section's position were mistaken for an enemy attack on the harbour!

The East African campaign which brought E and F Sections (and later all of 25 Battery) to Port Sudan is now almost a forgotten chapter in the chronicles of the Second World War. In the mid-thirties the Italians had invaded Abyssinia (Ethiopia) and had overthrown the Emperor, Haile Selassie. Italy thus had an East African empire which threatened British Somaliland, Kenya, the Anglo-Egyptian Sudan and indeed Egypt itself. Italian intentions included the capture of British possessions in the area and the conquest of Egypt. On August 11th 1940 seventeen Italian infantry battalions supported by two artillery groups and armour attacked British Somaliland. The defence (which was composed of a battalion of the Black Watch, two Indian battalions, two East African battalions, the Somaliland Camel Corps and a minimal force of artillery) was forced to withdraw to enable an evacuation to be made to Aden. That evacuation began on August 15th and soon after the Duke of Aosta held Somaliland and had a force of 250,000 troops facing the British in the Sudan

In this situation the Commander-in-Chief, Middle East, General Wavell, prepared first for the defence of Sudan and then for an invasion of Italy's East African empire. For this strategy Port Sudan was vital and thus required the protection of anti-aircraft guns which was supplied by 25 Bty.

After the alarm on July 12th, which was presumably a reconnaissance aircraft, there was a spell of relative calm until August 1st when at 8.05 a m an Italian bomber dropped two bombs close to one of the gun positions. This aircraft had been mistaken

[2] The gun positions were on coral rock thus necessitating the blasting operations. False alarms were an occupational hazard for AA gunners.

for an RAF Blenheim flying out of the approved approach corridor for Port Sudan. The "corridor" basically was an approach from the south at under 10,000 feet and a Blenheim had been expected at 8.00 a m! In spite of the degree of surprise which the Italian crew unwittingly achieved they came close to being shot down as at least one shell burst close to their aeroplane. John Ormsby recalls watching the 'plane rocking violently as the shells bracketed it before it flew out of range.

When the Somaliland campaign got under way the remainder of 25 Battery were on their way to Port Sudan from Port Said to where they had moved briefly from Agami. F Section, under Lieutenant Harold McCullagh, were ordered to move from Port Sudan by sea to British Somaliland, but the orders were changed to send them instead to Aden, across the Gulf from Somaliland. August 2nd was spent loading their guns, vehicles, 2,400 rounds of ammunition and three days rations on board two ships. Lieutenant E H Babington, who paid the Master of the "*Akbar*", was shocked to learn that the ship could only make four knots. But for those who sailed on the "*Akbar*" there was an even greater shock in store. Never before had any of the Derrymen seen such filth and squalor as they saw on that ship, and the cockroaches must have been the world's biggest! The Greek master and the owners certainly had no concern for the health of the "*Akbar's*" normal passengers — Islamic pilgrims travelling to Mecca.

It took several days in convoy, under destroyer and air escort, to reach Aden where F Section were soon ready for action. In spite of regular alerts they did not open fire — Italian bombers operating in the Gulf seemed to drop their bombs out to sea — but they did fall foul of what can only be described as the peacetime army. One prewar Regimental Sergeant Major, on observing their uniforms, already much the worse for service in Sudan and the voyage to Aden, described them as a "disgrace to the British Army". Little did he know just what a reputation the regiment to which these "scruffs" belonged would build for itself in a short time.

H Section arrived in Port Sudan on August 19th taking over E Section's position. Ten days later G Section came, and with it Captain Sir Basil McFarland, Bt, and Battery HQ. G Section's guns and equipment were delivered on August 31st and three days later F Section returned to Port Sudan from Aden.

On September 7th all four sections went into action against two SM 79 bombers. These aircraft flew over the town from north to south and then returned, dropping eight bombs on the docks. Fortunately there were neither casualties nor damage. On their first run the bombers were engaged by F Section and then, as they returned, by F, H and G Sections. Two Gladiator fighters from the nearby RAF Station took off to intercept the raiders, and with friendly aircraft in the vicinity the guns were forced to cease fire. E and H Sections however fired two "pointer" bursts each to direct the fighters to the enemy machines.

Next day a convoy of five troopships and their escorts arrived in the harbour and at 11.45 a m enemy aircraft attacked. Once again fighters went up to intercept and 25 was unable to engage for fear of hitting the Gladiators, although F Section did loose off a salvo before the fighters closed in.

During September there was a steady stream of arrivals at Port Sudan as Wavell built up his forces for the planned attack. This activity continued into October, when there were several attacks by Italian bombers. The worst attack occurred on October 23rd when a raid by SM 79s killed twenty-seven Sudanese dockers and injured twenty-three others, some of them seriously. The last bomb fell close to E Section's position but the gunners suffered no casualties. Despite this horrific loss the Sudanese workforce were soon back to normal work, unloading the many ships which docked in the harbour.

The Battery had two VIP visitors at the end of the month with General Wavell, the C-in-C, inspecting H Section on October 30th. Mr Anthony Eden, then the Secretary of State for War, visited them the following day and inspected E and G Sections. Eden was in the Middle East on a fact-finding mission for the Prime Minister, Mr Churchill, and also visited other elements of 9th (Londonderry) Regiment at Alexandria. As he left Eden told Major Thompson "You are a lucky man to command such a fine battery".

Italian air activity against Port Sudan intensified during November, which also saw the beginning of the British counter-offensive. During the November attacks the Battery had their first confirmed destruction of an enemy aircraft in Sudan. This occurred on November 15th when two SM 79s attacked Port Sudan at 11.37 a m. One of the aircraft was hit and claimed as a "probable"; in fact it was destroyed and for this confirmation the Battery had an unexpected source — Rome Radio, which announced that one of their aircraft had not returned to base. The raid did cause some damage when cotton bales, piled up at the harbour since they could not be exported due to the war, were set on fire. Another pair of SM 79s came over on November 21st and were engaged by E, F and H Sections together with the guns of two warships in the harbour. RAF fighters again forced the guns to cease fire. No hits were claimed and the attackers had caused no real damage.

On November 6th British troops recaptured Gallabat in Sudan which had earlier been taken by the Italians. Next day G Section/25 Battery were ordered to move to defend the railhead at Atbara, a vital staging-post for men and supplies for one of the "pincers" of Wavell's attack on Aosta's forces. G Section arrived at Atbara to find that "most satisfactory gun and instrument emplacements had been prepared by Royal Engineers".[3] They were ready for action next day but for a week could see virtually nothing as the area was subjected to sandstorms which blinded them and sapped their energy. After a break of four days the sandstorms returned and raged for another three days. This was to be the pattern of G Section's life at this post, and although they were called out on alerts on numerous occasions they saw no action — only sandstorms!

During December there were eleven enemy air attacks on Port Sudan. These included a reconnaissance plane on December 8th which was engaged at almost 20,000 feet. At 3.45 a m on December 16th one aircraft bombed the town but its bombs failed to explode. Fifteen minutes before noon that day a flight of three bombers attacked and were engaged by both guns and fighters. One bomber was shot

[3] 25 Bty War Diary.

down by the guns and three of the crew of five were killed and the others injured. An hour later two bombers appeared and were subjected to fire from the guns. One bomber was hit and crashed into the Red Sea. Fifteen minutes later yet another three bombers arrived over Port Sudan and began to bomb not the harbour but 25 Battery's positions! One bomb exploded close to H Section's guns damaging the GL set and the gravity tank on No 2 gun. Another bomb fell on the cookhouse. Miraculously there were no casualties except as one Derryman wryly remarked "all our Christmas turkeys". In truth only one of the turkeys which had been kept alive beside the cookhouse was killed. The Battery Captain's car, J J Doherty's motorbike, and various other pieces of equipment also suffered splinter damage.[4]

The damage to the GL set and to No 2 gun was quickly repaired, the gravity tank being plugged with wood, and H Section was back in business for the next raid which came almost as repairs were completed at 2.00 p m. This attack was carried out by one aircraft which again homed in on 25 rather than the harbour or the airfield. The pilot's briefing had obviously been based on out of date intelligence for he attacked G Section's former position which had lain vacant since they moved to Atbara. Tragically however the bombs killed a number of Indian soldiers of the Bombay Sappers and Miners, who for some inexplicable reason had not taken cover when the alarm sounded. Another six Indian soldiers were injured.

There were no further raids that day, but at 2.45 next morning a "stand-to" was ordered when the RAF radar picked up an aircraft approaching Port Sudan which turned away when it was still some twenty miles distant. However at 9.00 a m a single raider was heard approaching. Visibility was poor with almost 100% cloud cover and the aircraft was seen only for a few seconds through a gap in the clouds as it bombed the south side of the harbour, setting a petrol and oil dump on fire. Other bombs fell harmlessly in the sea. The guns did not fire due to the poor visibility and the fact that fighters were up. There were a further three raids that day but none had the effects of the first attack or of the attacks on the previous day.

The *Regia Aeronautica* then appears to have taken a Christmas holiday for no further attacks were launched until January 5th 1941. It is possible however that for some of that time at least the bombers were prevented from operating by the weather conditions, dust and sandstorms not being particularly conducive to safe take-offs and landings. A single SM 79 was engaged over Port Sudan on January 5th, and ten days passed before the next raid which occurred at 8.40 p m on January 15th. Six bombs were dropped causing minor damage and no hits were claimed by the guns. A number of other radar "pick-ups" led to alerts that day but no action occurred.

On January 19th the British attack on Eritrea[5] was launched from Sudan. The remaining area of Sudan which had been held by the Italians was soon recaptured

[4] Billy Smyth recalls that the motorbike was destroyed seconds after his friend J J Doherty had parked it and joined him in a slit-trench beside the Command Post.

[5] Eritrea and Italian Somaliland had been Italian possessions for some fifty years and had provided the springboard for the annexation of Abyssinia — the Eritrean port of Massawa on the Red Sea being used to bring in supplies and personnel for the attack.

and Lieutenant-General Platt pushed his two division strong force towards the Asmara Plateau. At the same time Lieutenant-General Sir Alan Cunningham (an Irishman and the brother of the admiral commanding the Mediterranean Fleet) advanced north from Kenya.

G Section/25 Battery was now ordered to move to Kassala on the Border between Sudan and Eritrea, a post which had just been liberated from Italian control. The very journey to Kassala on January 22nd was a nightmare. It was "terrible country and but for the winch gear the guns could not have been pulled to Kassala".[6] When they did arrive there the gunners found that they had to sleep under bushes since, because of camouflage difficulties, they could not pitch their tents. That soon became the least of their problems for the area was infested with white ants whose marauding attentions were maddening.

As at Atbara, G Section saw alerts but no action and had to suffer heat, flies and those always present white ants. A sighting of an Italian aircraft was made on February 8th, but this turned out to be a Caproni 133 in RAF hands which landed at Kassala. This aircraft was a frequent visitor to Kassala and caused a few alerts but it appears to have been the only "enemy" aircraft which G Section saw during February.

With Aosta's forces in East Africa now on the retreat, relief was announced for 25 Battery at Port Sudan. That relief arrived on March 13th in the form of 23 HAA Battery of the Hong Kong and Singapore Royal Artillery (HKSRA) under which F Section had served in Aden. On March 15th G Section were ordered to return from Kassala. They went out of action that day and prepared to move next day when at 11.00 a m the temperature at Kassala was recorded at 152°F! At 8.00 p m they left Kassala and arrived back at Port Sudan on March 18th.

That temperature of 152°F at Kassala is an indication of just how unpleasant life could be in the Sudan. At Port Sudan itself temperatures regularly reached 140°F during the day. In the gun pits it could be even higher and at night the temperature drop was sufficient to encourage the guards to wear greatcoats to the surprise of navy personnel who would come ashore when the temperature was still in the 80s. The gunners lived in huts which had been built originally as accommodation for dock labourers.[7] When they went to sleep it was too warm to use blankets and most men would sleep virtually naked. This in fact could cause "gyppy tummy" which could be prevented by sleeping with a towel wrapped around the stomach — an old soldier's trick which was passed on to John Ormsby by J J Doherty, who was one of the few old soldiers actually prepared to give such advice to the younger and less experienced gunners. The climate in Sudan caused many health problems — one of which was the gunner who drank lemonade to quench his thirst and then found his stomach distending so painfully that he was unfit for duty! In consequence, small groups of men would be despatched for rest and recuperation ninety miles away to the hill station at Gebeit, where at 3,000 feet above sea level, it was much cooler. There they admired the huge regimental badges set into the hillside which in fact remain to this

[6] 25 Bty War Diary.

[7] In typical army fashion these hot-houses were known as "igloos" by the gunners.

day. A lot of interest was taken in the Royal Irish Fusiliers' badge, not least because their fellow gunner "J J" had helped construct it seven years before.

There were mercifully few casualties in Sudan. It could have been otherwise, for the Italian bombers concentrated their efforts on the gun positions on many occasions. The most serious casualty in their time in Sudan was caused by an enemy bomb which had failed to explode on landing and which was detonated as a convoy drove along the road from Port Sudan to Suakim. Caught in the blast and hurled down the road was the Don R (Despatch Rider) escorting that convoy. J J Doherty was seriously injured by shrapnel which required several weeks in hospital and later probably contributed to his early death from cancer of the intestines (where some of the shrapnel had lodged). Detonation of the bomb was caused by another member of 25 Battery who fired a rifle shot at it!

But before 25 Battery could say goodbye to Sudan they were involved in one last major alert. This was an alert with a difference, for an attack on Port Sudan by fast surface units of the Italian navy was expected, and the guns were deployed to engage either sea or air targets or both. G Section in fact moved almost on to the beach together with B Section of 23 Bty HKSRA and a Coastal Battery, while E, F and H were instructed to engage any attacking aircraft or, in the event of a seaborne attack only, any naval targets on which they could bear.

On April 2nd information was received that Italian destroyers had sailed from Massawa and would attack Port Sudan. Early next morning Fleet Air Arm Swordfish aircraft lurched off from Port Sudan carrying torpedoes. Explosions were heard out to sea and during the morning there was a constant shuttle of Swordfish, Wellesley and Blenheim aircraft from Port Sudan. It was later learned that the aircraft had intercepted the destroyers and sunk them. Six Italian ships went down to the bomber and torpedo planes in a gallant but ill-fated venture; many of the survivors were later landed at Port Sudan to spend the rest of the war as prisoners.

That evening, the alert over, 25 Battery stood down and prepared for their departure from Port Sudan. Ahead, they were assured, lay a spell in a rest camp before resuming their duties in the Alexandria defence area.

CHAPTER TWO

IN THE HOTHOUSE

The deployment of 25 Bty to El Daba and then to Port Sudan, with detachments to Aden and Kassala, was directly attributable to the dearth of AA resources in the Middle East at that time. All eight guns of the Battery were together as the AA defence of Port Sudan for only a relatively short period, and it was fortunate that the Italian air force in the region was as severely stretched as the Royal Artillery's anti-aircraft resources.

Service in the Sudan had many unpleasant aspects, not the least of which was the sub-tropical heat, which seemed almost to have a physical presence. "Prickly heat" afflicted everyone and the healthy suntan of Egypt began to drain away from their bodies which took on a washed-out white pallor — a condition familiar to old India hands in the monsoon season. Men lost weight and everyone seemed to have grown older although for much of the time features were hidden beneath their tropical pith-helmets or topees and the dark glasses which were needed to cut down the glare from the dazzling white coral of the harbour area.

That harbour was essential for the East African campaign because it was the only one between Suez and Massawa in Italian-held Eritrea. At any time the port was crammed with ships bringing in men and equipment for the campaign. Its defence against air attack was the responsibility of 25 Bty. Battery HQ was for a time located in the Mudirya or Commissioner's office and the GOR was also there.

The local population consisted of Sudanese, Greeks and British, the latter in particular making the soldiers welcome. There was an area of several streets of bungalows for the British residents, and a similar Greek colony. Hadendoa tribesmen — the legendary "fuzzie-wuzzies" — made up the native population. A proud people — had they not had the unique distinction of having broken a square of British infantry? — they were vastly different from the natives of Egypt. No stench of garlic and sweat pervaded Port Sudan, nor were there the eternal cries of "baksheesh" as in Egypt; it was like another world.

The local British residents organised tea parties for the gunners and a club known as the Pelican Club; both of these facilities were extremely welcome. Greek-run eating-houses were also frequented; usually this happened when rations had been cut yet again due to another influx of fresh troops and then only for a plateful of rice pudding! — a gunner's pay could buy little more.

For entertainment there was a cinema; not a plush picture palace but an open-air facility which was noteworthy for the hedgerow growing in it which served as a definition of sorts between stalls and gallery! Although the films were old and the projector would frequently break down it was at least a pleasant way to spend an off-duty evening.

G section of course spent four months away from the Battery at Atbara and Kassala. For them there was no diversion at all as they soldiered in some of the most inhospitable country in the world without any action to relieve the tedium. Their entertainment was provided by themselves, with men like George Lapsley preventing their comrades from dying of boredom. Lapsley's singing and his humour were invaluable to morale.

When "Two-Five" left Port Sudan for the journey back to "Alex" there was no-one who was sorry at having to say goodbye. Perhaps there might have been second thoughts had they but known what the next six months held in store for them!

CHAPTER THREE

THE BENGHAZI STAKES

When, at the beginning of April 1941, 25 Battery were ordered to pack up and leave Port Sudan there were few regrets at saying goodbye to this hothouse climate and returning to Alexandria. Loading of the Battery's equipment and embarkation of personnel was complete by April 8th and the following day the SS "*Khedive Ishmael*" set sail up the Red Sea with the main party of the Battery's men, arriving at Suez two days later. They disembarked and entrained on April 12th, arrived at Cairo that evening, travelling onward to Alexandria after midnight. They arrived back in "Alex" at 6.00 a m on Easter Sunday morning looking forward to the promised spell in a rest camp before resuming duties in the "Alex" area.

While they had been moving however so too had Rommel! The Desert Fox was now at the Egyptian border, and GHQ were rushing reinforcements up the desert to halt that advance. Imagine therefore the chagrin of 25 Battery's men when they were given two hours leave and ordered to be ready to move out again before noon. They had come back from the Sudan in need of a rest — would they ever get one? Their clothing was showing the effects of the Sudanese climate — bleached to an indefinable colour, often tattered and torn, it could hardly be replaced in a matter of hours — although some gunners did manage to get a change of uniform by swopping with colleagues from 24 Battery who could hand in the old kit for replacement.

At 11.30 that morning they left Alexandria and ten hours later arrived at Mersa Matruh, where they found that no arrangements had been made for their arrival! Empty houses in the village — which had been a summer resort for better-off Egyptians before the war — were used as temporary billets that night. Equipment began to arrive the next day and on April 15th their guns and transport which they had last seen at Suez reached them. That night, although they did not know it then, their home city was bombed, a parachute mine dropped by a single German medium bomber falling on Messines Park.

The positions which they took over had been designed for 3 inch guns and required much hard work to make them suitable for 3·7s. Nonetheless they were ready for action on April 17th and also took over responsibility for the Gun Operations Room of Matruh Fortress four days later. The dugouts which they took over from the Egyptian Army were another story altogether. That so many bugs existed in one place was unbelievable. But they did, and all sorts of remedies were tried to kill off the bug population, including coating the interior walls of the dugouts with a paraffin mixture. One soldier coated himself with a similar mixture only to find the cure worse than the problem!

My father only ever mentioned one warlike incident to me and that arose from a question he asked me about a model aircraft I had just built when I was about 13 years old. The model was of a Heinkel He111, an aircraft which he referred to as a "Trouble-One". This nickname was the result of a belief that these aircraft were equipped with a device that could "home" on the signals of hostile radar. As the alarm sounded at 6.30 p m on April 25th my father was in the presence of a Second-Lieutenant who had joined the Battery from England only days before. He identified the approaching aircraft as a Heinkel for the young subaltern and suggested that its target might be the radar. The officer on being told the reason for this theory dismissed it as nonsense claiming "the Germans don't even have radar" (sic). When the pair emerged from cover after an engagement in which several bombs exploded and 14 rounds were fired by 'E' and 'F' Sections it was to find that a bomb had landed near the GL set, the subsequent explosion injuring six men, one of whom, Gunner Tommy Porter, died of stomach wounds eight hours later. Gunner Porter, one of a number of Donegal men in 25 Battery, was buried that afternoon at Mersa Matruh British Military Cemetery, the service being conducted by a Presbyterian Chaplain, Padre Henderson who had studied at Magee College in Derry. Padre Henderson had been by Tommy Porter's side as he lay dying in the Field Hospital.

Incidentally the Germans did have a radar homing device which could be fitted to an aircraft. Refinements of this device were to cause terrible losses in Bomber Command later in the war, when night-fighters were able to "home" on the tail-mounted radar of RAF bombers. This latter radar — codenamed "Monica" — was designed to combat night-fighters; instead it helped them!

Major Thompson, 25's Commander, was appointed AA Defence Commander, Matruh Fortress and issued standing orders for the defences on April 20th. Except when training the Section at No 3 Position was always to maintain one gun with a round in the loading tray ready to fire as an alarm signal for the area. No doubt bearing in mind the experience in Sudan when the gunsites themselves had been targeted by the enemy an instruction was issued not to fire on single reconnaissance aircraft flying at more than 20,000 feet.

Matruh was the strongpoint at which it was hoped to stop Rommel's drive into Egypt — hence the term Matruh Fortress — and the Battery were to remain until mid-May when two sections would move further up the "blue" to Sidi Barrani and the remainder would redeploy in the general area of Mersa Matruh.

Until then however they were in action at Matruh on nine occasions. The intruders were both German and Italian, and bombs were dropped on Matruh Fortress on four occasions. The Battery suffered no casualties or damage however on any of these occasions and put up 172 rounds in defence of the area. On the day that H and F Sections moved off on their journey to Sidi Barrani three of the four bombing raids occurred. The first occurred in darkness and the intruders — two Junkers Ju88s — were not engaged as the searchlights failed to pick them up. Another attack before dawn was similarly unopposed but a He111 at 6.30 a m was fired on with a total of 36 rounds from all eight guns of the Battery.

F and H Sections arrived at Sidi Barrani the following day, with some damage to the guns due to the nature of the roads. They were however ready for action later that same day, guarding a supply dump south of Barrani. No proper gunpits were constructed as they were to be ready to move again at a few hours notice.

At Barrani the gunners could actually hear the German and Italian aircraft taking off from their forward landing-strips in the Sollum area. Minutes later those aircraft could be engaged on a strafing attack on the guns.

Fortunately 25 Battery suffered no serious casualties during those attacks. It could so easily have been otherwise. John Ormsby recalls the occasion he was shaving in a dug-out when an alert began. Abandoning his shaving-tackle he dashed from the dug-out and jumped into the nearby gunpit, swinging on the barrel of the 3·7 just as Sergeant Bill Loughlin gave the order to fire. The result was burst ear-drums for Ormsby as he dropped into the pit. That discomfort however proved to have been the price of survival for when the 'all-clear' came and he was able to return to the dug-out it was to find that one of the attacking fighters had peppered it with cannon shells. His kit bag and other items had been shot through and to this day John Ormsby has his scorched AB64 (Soldier's Pay Book) to testify to his own "close shave".

Another strafing attack saw the attacking German fighter shot down by an RAF plane. The German baled out and found himself drifting down close to 25's position which only moments before he had been shooting up. Needless to say there were some very angry gunners, especially as the attack had occurred at a mealtime, and one man decided to teach the German a lesson. This gunner drove out towards the descending pilot and hit him as he landed, rendering him semi-conscious. He then proceeded to tie the German's parachute rigging to the rear bumper of the vehicle! Dragging the German over rock and grit would surely have killed him! However as the gunner climbed back into his vehicle to begin his "lesson" he was forced to stop by a motorcyclist. J J Doherty then cut the German free and escorted him safely back to the Battery's positions. To show his gratitude the *Luftwaffe* man handed his enemy-turned-saviour the only item of value in his possession — a German stamp bearing the head of Adolf Hitler! The stamp remained a treasured souvenir for J J although he would never explain how it came into his possession other than to say "a German pilot gave it to me". (The full story came to light when a casual mention of "the stamp" was made to a 25 Bty veteran during the research for this book.)

Even during strafing attacks there could be moments of high humour — as on the occasion when Jim Dinsmore was seen running for cover with a messtin full of rice on his head! Today Dinsmore, who "kept the Battery entertained" throughout the war according to Sir Basil McFarland, avers that the rice on his scalp caused his subsequent baldness!

Meanwhile G Section and Rear HQ had moved to a position east of Qasaba and E Section to Sidi Haneish airfield — both in the Mersa Matruh area. On May 19th F and H Sections together with Advance HQ were ordered back to the Matruh area with F Section moving to the airfield at Gerwala, Advance HQ joining Rear HQ to reform a full Battery HQ at Qasaba and H Section moving to Sidi Haneish. All were ready for action that evening having been on the move from 7.00 a m.

G Section at Qasaba picked up a Junkers Ju88 at 8.00 a m on May 31st but it was out of range for the guns. However it was reported to GOR and fighters shot it down west of Matruh.

They were soon to move again. After a relatively quiet few weeks, interrupted only by an inspection by the Brigadier, Royal Artillery (BRA) of Western Desert

Aden. Guns of F Section, 25 HAA Battery ready for action at Aden, 1940.
(Photo: Sir Basil McFarland's Collection)

"Up the Blue", June 1941. A crew of 25 Battery dash to their guns at Sidi Barrani.
(Photo: Imperial War Museum)

Force, they were ordered, as a Battery, to Sidi Barrani where they were ready for action on June 13th. Two days later orders were received that from 6.00 a m on June 16th E and G Sections with an Advance HQ were to be on 1½ hours notice to move to Sollum. When those orders were received Sollum was in enemy hands!

Wavell launched Operation "Battleaxe" that day. Spurred on, against his own judgement, by Churchill who wanted a military return for the risks he had taken with the "Tiger" convoy, the Commander-in-Chief ordered XIII Corps under Lieutenant-General Sir Noel Beresford-Peirse (another Irish general!) to attack through the Halfaya Pass, to move on to relieve Tobruk and, if possible, to move up to the Derna-Mechili area.

Whilst the two sections waited for the order to advance to Sollum, an order that would never come, the Battery was alerted at 1.20 p m to approaching aircraft. Three fighters flew in low towards their positions, one being under attack from the others. The attackers were Messerschmitt Bf 109s, and they shot down the RAF Hurricane reconnaissance aircraft they had been pursuing. As the Hurricane pilot ran from his wrecked machine the two German fighters made another strafing run and the RAF man fell, mortally wounded. Although the fighters were too low for the 3·7s to engage, some of the gunners opened fire with the close defence Bren guns. Percy Hamilton got a burst in down the fuselage of one Messerschmitt which crashed in flames. The other flew off and 'G' Section fired two 3·7 rounds as it climbed away. The Hurricane pilot who was called McFaul died in the arms of one of the "Derry Boys". As he lay dying he asked Captain Sir Basil McFarland if he could be buried at Alexandria. It was not a request that could be granted as bodies, for hygiene reasons, had to be interred as soon as possible and so the airman was buried near the site of the crash.

Less than five hours later a further five Bf 109s attacked the area but were forced to turn away as the guns opened up on them.

Although Wavell's attack enjoyed some initial tactical success, with Capuzzo falling to 22nd Guards Brigade and 4th Armoured Brigade, "BATTLEAXE" was soon blunted by the German anti-tank guns and smashed by superior tactics. German and Italian forces then pushed forward through 'Hellfire' Pass and towards Barrani on the coast.

With Rommel now inside Egypt E and F Sections were ordered to evacuate their positions and to rendezvous at Kilo 127 on the road to Matruh. When they reached the rendezvous they were ordered forward again to the positions they had left that morning! Having done so they received yet another order — to be at Kilo 127 at 0800 next morning.

When E and G Sections did move out, finally, on June 19th they left F and H Sections in Sidi Barrani with three days water and seven days food supply. That afternoon E and G took up positions around LG (landing ground) 07. The GL radar followed these sections to LG 07 two days later.

On June 27th LG 07 was evacuated by the RAF and the guns were told to prepare to move. The move began on June 29th but was delayed due to shortage of transport, not being completed until late the following night when the two sections had fallen back to the Fuka area, east of Mersa Matruh. Here they established their positions, although Battery HQ, originally sited near Abu Ahagag railway station, had to be moved due to severe daily sandstorms.

F and H Sections were still at Barrani, over fifty miles away as the crow flies, where they came under attack on July 4th by over twenty enemy fighters strafing them and the Landing Ground which they defended. At Fuka E and G opened fire on an SM79 at 20,000 feet on July 9th expending 19 rounds while on July 13th H Section fired 8 rounds at three Bf 109s which attacked them at Barrani.

F and H moved from Barrani to Matruh on July 19th where they came under South African Command and took up positions on the golf links and on the eastern spit. H Section fired four rounds at an SM79 on the morning of July 23rd. They so impressed the South Africans with their efficiency that the South African Official War History contains a special mention of 25 Battery.[1]

August was a relatively quiet month for E and G Sections but the same was hardly true for F and H. Billy Smyth recalls that there was hardly a night without an alert at Matruh. Unfortunately the Battery War Diary does not record F and H's experiences, although John Kershaw did record that approximately 2,000[2] rounds were fired during the month — mostly at night. Of these over 400 were fired on the night of August 4/5th. Bombs fell near them virtually every night but no casualties occurred in the Battery.

Both sides in the desert war were now rebuilding their forces ready for the next major phase of what the soldiers were wryly describing as the "Benghazi Stakes". Wavell had of course been replaced by Auchinleck and XIII Corps was finding a sister corps being formed alongside it — XXX — to create, from September, the most famous British army of all time — Eighth Army — and the *Afrika Korps* became part of *Panzer Gruppe Afrika*.[3] In the meantime 25 Battery had its own re-organisation as E, F, G and H Sections became A, B, C and D Sections respectively. (26 Battery's I, J, K and L Sections also became A, B, C and D. 24 Battery's sections of course remained unchanged.)

Severe sandstorms took their toll of men and equipment. Many soldiers found themselves suffering from desert sores. With water strictly rationed, bathing or showering was out of the question and every effort was made to conserve supplies. The result was that no matter how clean a man tried to be, his body became coated in a fine layer of gritty sand. Any cut, however slight, refused to heal in such circumstances and this was exacerbated by the shortage of fresh vegetables and fruit in their diet.

Positions were moved in an attempt to avoid the worst of the sandstorms and in the midst of all this a number of engagements took place — principally against recce aircraft. The last week of August saw the Battery move yet again to Sidi Barrani and they built new gun positions there before occupying them on August 31st.

[1] I am indebted to Mr James Soutar (formerly 25 Bty) for this information.

[2] TWO-FIVE — Experiences of a HAA Battery" (see Bibliography).

[3] All Axis forces in North Africa became *PANZERARMEE AFRIKA* with *Afrika Korps* continuing as part of that force. Towards the end of the campaign the overall Axis command was Army Group, Africa.

Two days later the aerodrome which they were defending, LG 05, was strafed by Fiat G 50 fighters at noon. Almost twenty of the Italian aircraft attacked and A and C Sections fired 29 rounds. The other sections were unable to engage as RAF fighters closed for combat. Seven G 50s were shot down. One crashed near to B and D's positions but in spite of attempts to rescue him the pilot perished as his aircraft caught fire. The death was even more poignant as somehow his paybook and wallet were flung clear of the wreckage. Some of the gunners believed that the pilot threw these out himself. The wallet, containing photos of the Italian's wife and four children, and his paybook were passed to the Red Cross.

On September 5th, also at noon, six fighters attacked LG 05. A and C fired a barrage of 14 rounds at 4,000 feet which caused the attack to be broken off, while B fired one round and nine were put up by D. Four hours later B and D fired 18 rounds at a recce plane at 18,000 feet.

The fighters came in again, twice, on September 7th. Just before 8.00 a m over twenty G 50s strafed LGs 02, 03 and 05 killing or wounding twenty-five men — among the wounded a gunner from A Section. A, C and B fired a total of 52 rounds but D was unable to engage. In mid-afternoon B fired seven rounds at six unidentified medium bombers of the RAF flying out of the approved corridor. Then at 5.27 p m six Bf 109s carried out a strafing attack. This time D was the only section able to engage and fired 7 rounds.

An Italian reconnaissance aircraft — an SM 79 — which came over at 2.00 p m the following day was engaged by all four sections. The aircraft, flying at 18,000 feet was seen to be hit by some of the 38 rounds fired and 22nd Guards Brigade HQ confirmed that it crashed into the sea. Inspired no doubt by this success 25's guns fired 97 rounds at a German Heinkel He 111 at 22,000 feet just twenty minutes later. No hits were claimed.

A Messerschmitt Bf 110 was seen at 21,000 feet on September 11th and 15 rounds were fired — again without effect.

Luftwaffe Bf 109s were engaged at 7.00 p m two days later. Flying at 20,000 feet the fighters had 20 rounds fired by them by B and D Sections.

Next morning at 10.00 a m all four sections fired a total of 48 rounds at a Ju 88. At noon the Battery was ordered to be ready to move. Eight hours later the sections were ordered to move at once independently and to rendezvous at Kilo 44 on the road to Matruh. A German armoured column had penetrated the front to the south and was heading northeast towards the coast — a move that would have cut off 25 Battery and put them in "the bag" as prisoners of war. All units were ordered to prepare to withdraw on receipt of the codeword "Bicycle". 25 Battery received the "Bicycle" code at 10.00 p m and in a frantic two hours were on the road to Kilo 44 where they laagered with vehicles dispersed and Brenguns at the ready. But the Germans had halted twenty miles south of Barrani and had then withdrawn when their petrol supplies were destroyed.

So 25 Battery received movement orders after ten hours dispersed at Kilo 44: go back and re-occupy the Barrani positions. This they did on September 16th. On the 17th they engaged a recce aircraft and two days later were ordered to hand-over the positions to 291 HAA Battery. During the hand-over a He111 was engaged by all four sections and 26 rounds were fired. On September 25th the GOC XIII Corps inspected A and C Sections and next day the hand-over to 291 Battery was completed.

There was a welcome day free to enjoy bathing in "the Med" and then on September 30th they left Barrani by road for the railhead at Matruh where they entrained on October 1st. When they arrived at Camp 17, Tahag (in the Nile Delta) next day it was the first time the whole Battery had been together in one camp since December 1939. Relaxation in the form of PT and sports followed until on October 6th they began to move to positions at Aboukir.

The take-over was complete by the 8th with Battery HQ at Malta Camp, B and C Sections at No 14 Position (Canopus) and A and D Sections at No 15 Position (El Burg). Administratively the Battery now came under 9th Regiment once again. Their wandering was over — they had returned to the fold but the experience they had gained in those long days in the desert was to prove to be invaluable indeed in the months that lay ahead.

CHAPTER FOUR

"THE BLUE"

The war in North Africa has been a favourite topic with the film industry in the years since 1945 and the deserts of that region have been depicted as a romantic area full of thousands of square miles of golden rolling sands — almost a vast European beach. In the way of popular art this image has come to be accepted as authentic by so many people who have never been near the scenes of the campaigns of 1940 to 1943. To the veteran who served in the desert the reality of the "Blue", as it was known, was totally different. The reality and the popular image are about as far apart as it is possible to be.

The truth is that the desert was a harsh, hostile and alien environment. Rather than being soft golden sand its surface was a harder limestone grit formed over centuries of wind erosion of the rock which forms much of Egypt and Libya. Each of these countries does possess both golden sand and rolling dunes in the southern sand seas, but the desert war was fought over a relatively narrow area of northern Egypt and Libya. In this region the surface was largely grit and rock and the colour which pre-dominated was a dirty mix of grey and brown, while towards the coast there was a strip of fertile land which contained most of the area's settlements and communi-cations. The southern sand sea was difficult terrain for any type of vehicle, and although some troops, such as the Long Range Desert Group and the Special Air Service, travelled through it the area was avoided by most combatants.

The desert was blazing hot during the day; literally hot enough to fry an egg, if you had one, on any exposed piece of metal! By contrast it could be freezing cold at night and, when there was no moon, so dark that it was impossible to find one's way over even the shortest distance.

In the "Blue" there was no diversion at all from the realities of the climate and the war. No-one in 25 Bty remembers any ENSA entertainment parties in the "Blue"; there were no cinemas and no food other than the never-changing army rations. Water was rationed strictly. 25 Bty had an issue of one gallon per man per day but, of this, half went to the cookhouse leaving the individual soldier with a half gallon to wash, shave, brew tea and keep his clothes clean. And he was lucky if he got the half gallon for inevitably water was lost through spillage and evaporation; it was not even pleasant to the taste as it had been "seasoned", in John Kershaw's words, with salt and chlorine, but it was all that there was. Occasionally there might be a ration of beer but it was so occasional as to be worthy of memory almost half of a century later. Boredom was often the greatest problem, recalls Sonny Babington, due to periods with little enemy action and nothing to do but watch the sky and wait, and wait.

In Part 2, Chapter 3 the affliction of desert sores was discussed and it was noted that this was exacerbated by the coating of grit which was constantly on the soldiers' skins. John Ormsby remembers that, when the guns were close to the sea at Sidi Barrani, this was relieved by an issue of saltwater soap and one gun at a time would be "stood down" so that its crew could bathe in the Med and remove the coat of dust. Even so it was not long before each soldier was again coated with that awful grit as he began to sweat once more and the dust stuck to his body. Desert sores remained a constant problem which was not helped by the almost total absence of fresh fruit and vegetables from their diet. Sores could be covered with bandages, but even these were in short supply, or coated with gentian violet when they were on areas such as lips. Some soldiers found that gentian violet helped while others reported no improvement with it. One gunner found that a spillage of "Brasso" on some of his sores actually helped to clear them up! Logic however dictated that "Brasso" should have worsened his condition and no-one else in the Battery was prepared to give this "cure" a try.

Cuts from shaving could almost be guaranteed to become desert sores yet shaving remained "*de rigueur*" as beards would have prevented anti-gas respirators from fitting snugly, thus allowing the face to be burned. Beards could also harbour lice and other undesirable companions! Apart from this, shaving had a morale effect in that it helped personal pride in appearance. Thus it was that only very rarely did a day pass without a shave even if the soldier had to use the remains of his mug of tea in place of water! Nothing was ever wasted in the desert.

Ultimately the only real answer for someone with a bad case of desert sores was to evacuate him to a base hospital for treatment, and at one stage during the summer of 1941 25 Bty had almost a quarter of its strength in hospital or convalescing from

The reality of life in the Western Desert. Sergeants Eric Wilson and Fred Holman make a shelter of sorts from the awning of a wrecked Italian lorry. September 1941.

(Photo: Jim Soutar)

Lt A R McClean shaving at Sidi Barrani, September 1941.

(Photo: Jim Soutar)

desert sores. That represented a total of more than eighty officers and men. The casualties included Sir Basil McFarland who, typically, was more concerned with the possible loss of his "Derry Boys" than with his own condition. In his diary Sir Basil noted that whenever a gunner was discharged from hospital or convalescent centre he went to the Royal Artillery Base Depot from which he was likely to be posted to the first vacant place in any gunner regiment rather than back to 25 Bty. With the rate of admissions to hospital from desert sores as high as it was this policy meant that there was a very real danger of "Two Five" losing its Derry identity and thus Sir Basil did all in his power to have his men returned to own Battery. He was largely successful in this although not every man came "home".

There could be little doubt that almost a year of service in the Sudan followed by this posting to the desert (not forgetting that half of the Battery had gone to Sudan from El Daba in the desert anyway) had taken a great toll of the health of 25's personnel. Senior officers who inspected the Battery at Matruh in late August realised that the men were worn out and would have to be relieved. Even so they were sent up to Sidi Barrani again in September and were fortunate not to have been captured there. (see Part 2, Chapter 3).

And then there were the flies! As if the grit and the desert sores were not enough the flies were added as if a series of plagues was being visited upon them. Those desert flies were no ordinary flies. The late John Kershaw likened them to the fanatical hordes of the Mahdi who, whenever one of their number was killed, came on in even greater numbers and with increasing ferocity. They were of course attracted to food, and mealtimes were a battle of wits with the flies. A popular tactic used to combat them at mealtimes was to cover the food with an extra enamel plate and then to surreptitiously lift this plate and grab a mouthful of food. But those flies were old campaigners and as soon as a cover was lifted they swooped like a *Geschwader* of Stukas. Human blood attracted them too and they soon found any desert sore which was not covered. And they feasted on the bodies of the dead; they were one of the reasons why the dead were buried so quickly in the desert. In his book "War in the Desert" James Lucas notes that the biggest difference a desert veteran would find now in his old battleground would be the comparative absence of flies, for of course today there are no armies in the desert with all the attractions for flies, especially the corpses.

The desert sore affliction did not end with the North African campaign. Many veterans found that they had health problems in the post-war days attributable to their service in the desert. Among these were dental problems, and Bertie Cuthbert recalls seeking treatment from a dentist shortly after the war who told him that his problems had been caused by his diet in the desert! In later years many men have found their health breaking down as a result of those days in the "Blue" and numbers of war pensions have been awarded because of this.

The army imposed very strict hygiene regulations even in the desert. Whenever a unit was static for any time latrines had to be constructed and these were normally built as a wooden box over a latrine trench surrounded by a hessian screen although there were no internal divisions. While a unit was on the move hygiene was still an important consideration and it was woe betide any soldier who tried to perform his natural functions without "taking his shovel for a walk".

Urinals were made from petrol tins from which the bottom had been cut; the tin was then sunk into the sand and another tin placed at an angle on top thus producing what was known as the "Desert Rose". Frequent treatment of latrines and urinals with chemicals, usually chloride, was carried out to reduce the fly population and thereby cut down the risk of flyborne diseases. Compared with either the German or Italian armies dysentery was relatively rare in the British army, thanks largely to these measures. The Axis armies also suffered from at least two outbreaks of diphtheria which did not afflict our forces at all.

Nor did the British army suffer from vermin as did the Axis forces — even though at places like Matruh British troops were alongside Egyptian soldiers whose standards of cleanliness were almost non-existent. When 25 Bty arrived at Matruh in April 1941 they were at first billetted in abandoned houses before moving into prepared dug-outs at the gunsites. But they had taken the sites over from Egyptian troops and found that the dug-outs were already inhabited! The problem was cured by flooding the dug-out floors with petrol and partially blocking the entrances before igniting the petrol. The scorched interior that was left proved to be remarkably free of bugs!

Perhaps the largest part of the reason for the British soldier having fewer problems in this sphere than his opponent lay in the anti-louse powder which was issued in the British units. There were also frequent hygiene checks carried out by the medical services and a number of these are noted in the War Diaries of batteries of the Ninth.

Clothing was kept clean by "dhobying" (laundering) with petrol or sand or even water. The life of khaki drill uniforms was however quite short in the desert with the conditions taking a rapid toll of the fabric, which was bleached by the sun, worn away by the grit, and rotted by sweat.

One of the biggest curses of the desert was the sandstorm. The sandstorm can occur at almost any time of the year — Ninth were greeted by one at Alex in November '39 — but the storms which occur during the spring have an especial quality of awfulness. During April and May when the warm wind blows up from the centre of Africa and lifts the sand and grit before it a hell on earth is created. At El Daba in May 1940 25 Bty had their first experience of this African affliction and their tents were blown here and there. Thereafter they were prepared for the *Khamseen* as the Arab calls the sandstorm. Not only does it blow everything in front of it out of the way but it also makes a very good attempt to suck the breath out of any human in its path. Even the most oppressive and sultry day in Europe is but a pale imitation of the *Khamseen*, when breathing becomes hard labour, the dust gets everywhere, visibility becomes non-existent, and all anyone can do is sit with his back to the storm and wait for it to blow over. Of all the unpleasant aspects of soldiering in the desert the one which undoubtedly takes top position is the *Khamseen*.

25 Bty finally returned to the Alexandria area and the comparative luxury of brick-built dug-outs at Aboukir — a far cry from the desert where their rest at night

Italian war graves at Sidi Barrani, 1941.

(Photo: Jim Soutar)

25 HAA Battery hockey team 1941. Front row (L to R) William Johnston, Jim Canning, Miles Stott, Ernie McClure and Thomson Jamieson; rear row Percy Hamilton, Jack Lee, Herbie Payne, "Pop" Lapsley, Jim Smith and Walter Wray.

(Photo: Jim Canning)

was taken in a hole scooped out of the ground and covered, if they were lucky, with a groundsheet or blanket!

For all its cruelty, or perhaps because of it, the desert retained a special place in their memories. None of 25 Bty's men have ever forgotten the road from Fuka to Sidi Barrani — especially that awful stretch of about 15 miles which threatened to wreck suspensions and equipment and which hardly deserved the description of "road". The landmarks still spring to mind — the tree to the north of the road some 30 miles west of Matruh, and the YMCA south of the road at Kilo 94 where cigarettes and occasionally tinned fruit could be bought as well as tea — provided that the buyer supplied his own water. Yes, there was certainly something special about this ante-room to Hell that was called the "BLUE".

CHAPTER FIVE

SPORTING TRIUMPHS!

Sport has long played an important part in army life even on active service. At times when enemy activity seemed relaxed recreational diversions would be arranged, and sports fixtures of all types were included.

Soccer was always a popular sporting activity and in addition to a Regimental team which challenged other units (and provided on occasion a number of members for an Irish team for "international" fixtures) there were inter-Battery matches which provided some very memorable moments. A match between 24 and 25 Batteries in the Canal Zone on St Patrick's Day 1942 resulted in a 1-1 draw but non-partisan spectators commented that 25 Battery and its supporters had had the edge in the amount and "quality" of verbal abuse which had been hurled at the other side. It must be remembered that this was shortly after 25 had returned from eighteen months in the wilderness while 24 had been static at "Alex" during all that time. Typical of the insults which can be repeated was the description of 24 as the "'Alex' Home Guard" and the riposte that 25 were a bunch of "Desert Rats". It was fortuitous that a sandstorm blew up shortly after the match preventing the Batteries from continuing their duel in other fashions.

On other occasions matches would be organised within a Battery between Sections or between the officers and NCOs or the Gunners. Jim Canning kindly provided the following contemporary report of a match in which he himself played at Tahag. The report was written by the referee, Bruce Campbell, and the teams on this occasion were the Officers and Sergeants of 25 Battery.

1941

LOCAL DERBY AT TAHAG

SERGEANTS WIN ANNUAL FIXTURE

WORTHY CHARITIES BENEFIT

The huge crowd present at Celtic Park, Tahag, on Sunday were treated to a vigorous and dashing display of football, on the occasion of the annual 'blood' match between the Officers and Sergeants of 25 Battery. The Officers' team was strengthened specially by the inclusion of Lt. Col. Aitken, Lieut. Hale and RSM Worgan.

TEAMS	OFFICERS:-	Ogilvie, Aitken, McFarland, Harpur, Hale, Butler, Crooks, Worgan, McCullagh, Bolton, Sheddick.
	SERGEANTS:-	Lapsley, Loughlin, Gamble, Marsh, Lee, Brown, Thompson, Wray, Jones (BSM), Canning, Fleming.

Punctually twenty minutes late Referee (No Surrender) Campbell lined up the teams. After winning the toss, Captain Padre Crooks elected to play with the slight breeze. From the kick-off the Officers pressed strongly, and Butler, when nicely placed in front of an empty goal kicked himself smartly on the ankle and left the way clear for Gamble to save. Play shifted rapidly to the other end and Wray (hi) sent in a sizzler which was well held by Ogilvie. From his clearance a nice movement followed in which McCullagh, Bolton and Sheddick all participated, and which resulted in McCullagh putting a low shot well past Lapsley. Wray and Thompson now took up the running, but McFarland, interposing himself vigorously, changed their ideas. A further duel, in which Shamus Canning endeavoured to shoulder the Major, resulted in the Hibernian being flung some twenty yards.

A period of shooting round the visitors' goal now followed. Buck Jones made a desperate effort to get to a passing ball, while Badgie Fleming missed from a range of two feet. Lee had a bash from a distance outside the area, but the ball was safely gathered by the Cameron man. Wray and Thompson again essayed a breakaway, and Wray's shot beat the goalie all the way.

Hale, whose elated kicking, had been shortening Brown's life on the wing, was now brought in to the centre and play proceeded vigorously. Harpur tackled Canning in the Rugby fashion, while Butler again fell, by his own feet.

A penalty awarded to the Officers gave Hale a chance to slip a nice one past Lapsley, whose lumbago prevented him from playing a low game.

Sergeants now pressed in their turn, and Wray had hard luck with a stiff shot. McFarland came up, brushing aside all opposition, but Gamble cleared cooly.

Half time came with no alteration in the score

Officers, 2 Sergeants, 1.

On the resumption, the Sergeants' goalie was seen to be in the left wing, while Brown became the new custodian.

From the kick-off Sergeants attacked, and Fleming tested the Highlander with a first time effort. Padre Crooks showed his versatility at this juncture, playing on both wings practically simultaneously. Brown made a great save of a high curling shot and cleared well. Lee and Fleming took the ball to the other end and a perfect centre from Fleming was headed by the Major through his own goal! Marsh and Sheddick now had a struggle for the ball, but Marsh, though giving weight away, was successful. Sheddick appeared hurt, but was able to resume despite the attentions of Trainer Wilson. The stretcher bearers (Noisy Herriott and T.T. Hunter) who were quickly on the scene, were relieved to have no work.

On resumption Sheddick and Bolton indulged in a well planned movement, but the whistle was blown for offside. A few minutes later, the Referee throwing discretion and all chances of promotion to the wind, penalised the RSM for a nice bit of elbow work.

The play at this stage was especially vigorous and Curley Lapsley and the Padre had several tussles. Thompson, Wray and Fleming put a barrage up the visitors' goals and a well placed shot from Thompson enabled Fleming to get the decider.

McFarland made a magnificent run but collided with Bolton, who got much the worse of the encounter.

In the closing minutes play slackened off and the Sergeants ran off worthy winners.

Final Score: Officers 2 (McCullagh, Hale)
Sergeants 3 (Wray, McFarland, own goal, Fleming)

At an entertainment later, the Supporters' Club, the Official Photographer, and the Press, were entertained by the winning team.

STOP PRESS

It is learned that Lieutenants Butler and Harpur had to seek medical attention after the match. After treatment they were able to proceed home.

There were many sporting encounters between Batteries, some of them being remembered as being particularly rumbustious. 24 and 25 Bty players recall the especially fine efforts made against them by 26 Bty's Arthur McNinch and there were certainly times when the football matches took on all the significance of a civil war. The gunner's traditional allegiance to his Battery was heightened in 9th Regiment by local loyalties within those batteries, and the rivalry between the Derrymen and the Braid men was often noted.

There was also a Regimental team which played with the same gusto as the Battery teams which provided its players, and at Tripoli in 1943 a number of 9th's footballers played in an "international" as members of an Ireland team against England. Needless to say it was a real needle match. The score has not been recorded but I have been assured that Ireland was the better team.

The footballers were helped in their training from a number of sources; 25 Bty while in Port Sudan received a quantity of footballs from "Ireland's Saturday Night" the Northern Ireland weekly sporting paper.

Nor was football the only activity. Tug o' War, always a popular activity in the army, had many adherents and featured regularly on the Regiment's sports programmes while basketball leagues were also run off.

From the hothouse of Sudan to the winter cold of Italy and on to the flat lands of eastern England sport played a tremendous part in sustaining the esprit de corps of the Ninth and it is small wonder that whenever veterans of the Regiment come together they are more likely to ask "do you remember the match at Kasfareet?" than to talk of bombs and shells and sweat and suffering.

PART THREE

THE END OF THE BEGINNING

CHAPTER ONE

DARKEST HOUR

During the long dark days between June 1940 and June 1941 Britain, the Commonwealth and the Empire had stood alone against the power of the Axis. The German invasion of Russia brought a new ally in the summer of 1941, and in December of that year the United States of America finally entered the war, following the Japanese attack on Pearl Harbour. Although no longer alone Britain, with its allies, faced a dismal strategic situation across the globe as 1942 began. That situation was to become even worse as the year wore on but as if to prove that the darkest hour does come just before the dawn 1942 was also the year in which the tide of war started to turn in favour of the allies.

That turn began in June 1942 with the US naval victory in the Battle of Midway. By the end of the year the Russians had begun to throw back the German attack at Stalingrad, and in the desert Montgomery had fought and won at El Alamein and had forced the Axis armies back almost to Tripoli.

During all of this time the men who had faced the enemy longest were those of the British forces in North Africa. Many of them, including the "Ninth", had been in action since June 1940, at times wondering if they would ever see home again.

There were mixed fortunes for the forces in North Africa in January 1942. For the army there was success — Auchinleck's Operation "Crusader" launched in November 1941 had taken 8th Army deep into Cyrenaica. Benghazi had fallen, again, and Rommel had been forced back to a defensive line at El Agheila. Bardia and Halfaya, which had been by-passed in the advance, fell to XXX Corps, with 14,000 prisoners in the first week of the month. Auchinleck estimated the enemy strength at El Agheila as being in the order of 35,000 men, some one-third of what it had been six weeks before.

The RAF's Middle East Command had shared to an extent in the army's success but the capable *Luftwaffe* adversaries faced by them were inflicting severe losses. I/JG27[1] with its Bf 109s was superior to its opponents — indeed in its first six weeks in the desert I/JG27's pilots had destroyed 63 allied aircraft, which was almost as many as the entire Italian air force had done in eight months. Its most distinguished pilot was undoubtedly Hans Joachim Marseille whose tally was 150 Allied aircraft destroyed (including 17 in one day!) by the time of his accidental death in September 1942.

[1] I/JG27 — 1st *Gruppe* (Wing) of *Jagdgeschwader* 27 (a *Geschwader* consisted of from three to six *Gruppen*, depending on role). A fighter *Gruppe* had about forty aircraft.

For the Royal Navy matters were even worse. "*Valiant*" and "*Queen Elizabeth*", resting on the bed of Alexandria harbour, would be out of action for some months. Worse still however another battleship, *HMS* "*Barham*", had been sunk by torpedoes off Sollum on November 25th with the loss of almost 900 lives, and just over a week earlier *HMS* "*Ark Royal*" had finally succumbed to enemy attack off Gibraltar. Other units of the Mediterranean Fleet had been sunk or immobilised, and at the beginning of January the Alexandria-based Eastern Mediterranean Fleet possessed but three cruisers[2] and a handful of destroyers.

The weakening of the Fleet at "Alex", added to the fact that the land battle was now hundreds of miles away in Cyrenaica, contributed to the pressure easing again on the defenders of Alexandria harbour. In the weeks between the turn of the year and March 10th, when the last elements of 9th Regiment moved out of Alexandria to the Canal Zone, there were only eight actions against enemy aircraft over the city. Out at Aboukir 25 Battery was involved in only one of these actions, all of which were against single reconnaissance aircraft; no bombs were dropped in this period. 25's only action was on January 6th when the guns under 9th Regiment's command fired a total of 105 rounds from Alexandria and Aboukir. The positions in Alexandria fired 165 rounds against recce aircraft on four occasions during February. Added to this was a "stand-to" on February 9th but although aircraft were heard none entered the defended area.

A movement order for the relief of the Regiment, by 73rd HAA Regiment, was issued on February 19th. 209, 210 and 211 Batteries began to move in a few days later. 26 Battery's relief by 209 Battery was virtually complete on February 28th when the incoming unit had its first engagement, during which two of its guns were still manned by the Ballymena gunners of 26 Battery.

RHQ and 25 Battery moved to Tahag on March 4th and on March 10th the last elements of 24 Battery finally left Alexandria. To 24 Battery went the distinction of firing the Regiment's last shots from the harbour. In the early hours of March 3rd they engaged enemy aircraft heading for the Canal Zone, firing a total of 45 rounds. Over a two-hour period before dawn the next morning they fired 309 rounds at bombers attacking the harbour from a number of different directions.

Since June 1940 the gunners of 9th Regiment had engaged more than 100 enemy air-raids. They had fired over 22,000 rounds of 3·7 inch ammunition and, during its time with the Regiment, 5 LAA Battery had fired over 1,300 rounds of 40 mm ammunition. They had destroyed five enemy aircraft with at least four more probably destroyed or shared with RAF fighters, in addition to the number which were damaged. The airmen of the *Luftwaffe* and *Regia Aeronautica* had failed, for all their efforts, in their primary objective — to destroy or damage shipping in Alexandria harbour and to deny the use of that harbour to the Royal Navy. Not one ship had been sunk or seriously damaged by enemy air action during that time.

On its departure from Alexandria the Regiment received a number of tributes

[2] The Australian Government had also recalled *HMAS "SYDNEY"* to home waters because of the Japanese threat.

for the work it had done. Brigadier E W G Wilson, commanding British Anti-Aircraft, Middle East, wrote to Col Aitken:

> "On the occasion of the relief of 9th Regiment at Alexandria I wish to take the opportunity of expressing my appreciation of the good service it has so consistently rendered in the defence of the Naval base. It has had to endure long periods of inactivity but efficiency never suffered and the shooting has been quick and accurate. It has effectively turned away enemy aircraft and prevented damage to the vital areas. The Regiment has played a very important part in attaining this result".

In his letter Wilson also praised the excellent handover which was a "credit to all ranks".

General Auchinleck had also passed a letter received from General Loch, Overseas Inspector, AA and CD, who on a visit had been much impressed with ack-ack in the area and said that the Middle East gunners could teach many lessons to the gunners at home. "Like a puppy with two tails" was how one 24 Bty officer described their feelings on receiving such fulsome praise. Needless to say 24 Bty also felt that most of that praise was directed to them!

From Tahag Regimental HQ and 24 Battery moved to Tel-el-Kebir where 24 prepared and occupied two 4-gun sites. 25 Battery moved out from Tahag to Kasfareet where they relieved 232 HAA Battery and also took over two 4-gun sites. Finally 26 Battery moved to El Shatt, Suez, where they too took over two 4-gun sites. At Tel-el-Kebir 9th Regiment also assumed operational responsibility for 235 HAA Battery. This did not increase the regiment's strength since, yet again, 25 Battery was out of the fold, coming under the operational control of, first 42nd LAA Regiment and, later, 13th LAA Regiment.

In March 1942 the organisation of the Batteries was modified once again and "Troops" were created. Each Battery's A and B Sections became Right Troop while C and D Sections became Left Troop. The Troops were commanded by Lieutenants — it would be another year before Troop Commanders were made Captains.

On April 23rd Regimental HQ and 24 Battery moved to Port Said, where they took over positions from 232 HAA Battery there and at Port Fouad. Two days later they fired 14 rounds at enemy aircraft passing over the area in a southerly direction. May was less than four hours old when the Battery engaged enemy bombers carrying out dive-bombing and mine-laying attacks on the docks. One aircraft was illuminated by searchlights while the remainder were engaged by GL controlled barrage. A total of 292 rounds was fired. A further mine-laying raid on the northern extremity of the Western Breakwater at Port Said was fought off with a total of 31 rounds.

Between May 19th and 21st, Regimental HQ and 24 Battery were relieved by RHQ, 2nd HAA Regiment and 20 Battery and moved back to Tahag. 235 HAA Battery left 9th Regiment's control to move to Suez, where it relieved 26 Battery which then moved to Tel-el-Kebir. 25 Battery remained at Kasfareet.

Fayid aerodrome was the next destination for 24 Battery and they moved there on June 24th. Here they received their first mobile Mk III 3·7 inch guns, and also said goodbye to their second battery commander in just over six months. Major Lambert had moved to Regimental HQ in December 1941 as second-in-command and his place had been taken by Major E E Grant. Lambert had been promoted to Lieutenant-Colonel and posted as AA Defence Commander, Cairo, in May, and Grant was moved to RHQ in his place. In turn Grant was replaced as 24's Commander by Captain J R Stockman of 235 Battery who was appointed temporary-Major.

Fayid was a major operational bomber base from which aircraft of the Royal Air Force would launch long-range raids on Tripoli and Benghazi (yet again in enemy hands). By the summer of 1942 the RAF was operating four-engined Liberators and Halifaxes from Fayid, and veterans of 24 Battery remember a succession of accidents involving the Liberators during their time at the Fayid base. As for enemy action however, they opened fire only once, on July 5th, when they discharged 42 rounds at aircraft flying over the airfield.

On July 19th 24 Battery were relieved at Fayid by 253 HAA Battery and moved to Camp 36, Qasassin, drawing eight new mobile Mk III 3·7 inch quick-firing guns when they arrived there. Regimental HQ moved to Qasassin also in preparation for their next deployment which they expected to be with 8th Army in Western Desert. During their time at Fayid the Regiment had also commanded 195 HAA Battery and a composite battery — P Battery at Abu Sueir — made up from members of 24, 25 and 26 Batteries. 26 Battery joined RHQ and 24 Battery at Qasassin having moved from Tel-el-Kebir to Kasfareet in mid-June. They had also had a change of Battery Commander, with Major Haughton transferring to the British Mission to the Egyptian Army. His place was taken by Captain W Robertson-Smith of 25 Battery who was granted the acting rank of major. At Kasfareet they had, with 25 Battery, come under the command of 13 LAA Regiment for operational purposes.

Since leaving Alexandria 26 Battery had been in action on five occasions. At El Shatt they had fired 10 rounds at a recce aircraft on April 10th, and at Kasfareet on four successive nights they had fired a total of 184 heavy rounds plus 75 rounds of light machine-gun fire at enemy aircraft. During March and April they had carried out intensive training in mobile work and in desert exercises, as far south as St. Katherine's Monastery on the west shore of the Red Sea. On moving to Qasassin they also collected a complement of new mobile Mk III 3·7s.

With its experience in the desert 25 Battery needed little mobile training. Indeed it had remained at Kasfareet since the move from "Alex". During early July it had, along with 26 Battery, engaged enemy aircraft on successive nights from the 3rd to the 7th firing a total of 156 heavy rounds. A course on the use of 3·7s as field artillery was held at Kasfareet and on July 20th the Battery moved to rejoin the Regiment at Qasassin collecting new guns en route.

At Kasfareet 25 Battery had created a spectacle which proved to be a tremendous morale-booster for personnel in the area when they staged a "Twelfth"[3] demonstration complete with Orange Arch, a band parade and speeches. The Arch

[3] 12th July — the Anniversary of the Battle of the Boyne when King William III (Prince of Orange) defeated King James II and secured the throne of England.

The "Twelfth" at Kasfareet. 25 HAA Battery's memorable Arch.

(Photo: Mrs R Semple)

was a work of art which overshadowed the efforts of 24 and 26 Batteries in earlier years. It was built by a cross-section of the Battery although the art-work was that of Johnnie Kitson. When the big day dawned a band was assembled with drums made from petrol drums and other items and flutes fashioned from anything suitable. The Arch and parade attracted the attention of many personnel in the area, especially Irishmen, and a massive party ensued which was matched only by 25's St. Patrick's Day party earlier that year. Incidentally, it must have been the only "Twelfth" on record when one of the platform speeches was made "on behalf of the opposition" by a Catholic member of the Battery!

At Qasassin the regiment was placed in Middle East Reserve. In view of the situation in the desert everyone expected to move to 8th Army, which had stopped Rommel's advance on the Delta at a line stretching south from the railway station at El Alamein to the Qattara Depression. Axis armies now stood poised only 60 miles west of Alexandria, which had been declared an "open" city.

The events which brought Rommel so close to the heartland of Egypt began in January, when he had counter-attacked from the El Agheila position. Benghazi fell seven days later, on January 28th, and Derna was captured on February 4th. The advance slowed near Gazala on February 7th and a period of relative "stability" reigned as Rommel built up for the next offensive which he launched on May 26th. Operation "Venezia" saw German and Italian troops hit the Allied Line at Gazala, where it finally broke. After a tremendous battle at Bir Hakeim, Free French forces withdrew towards Egypt, as did the Commonwealth forces of 8th Army. A garrison was left in Tobruk but the Axis star was in the ascendancy. On June 16th an Alexandria-bound convoy (Operation "Vigorous") was forced to turn back to Gibraltar while most of a Malta-bound convoy went to the bottom of "the Med". Then on June 21st Rommel took Tobruk (and with it 5 Light AA Battery, previously part of 9th Regiment). For this he was promoted to Field-Marshal and permitted to strike on towards Egypt — which he duly did. Auchinleck took personal command of 8th Army, which retreated to Mersa Matruh. Then Matruh was abandoned and Auchinleck ordered his troops back to the El Alamein line.

CHAPTER TWO

THE HOLY LAND

The victorious *Afrika Korps* reached El Alamein on June 30th. In Tripoli Mussolini arrived by air with a white charger for his victory parade in Cairo, while tropical uniforms were shipped in for the German-Italian army when it reached the Delta, and a special medal was struck to mark the fall of Egypt. At General HQ in Cairo it was equally believed that the end was near. On Wednesday, July 1st, GHQ, Cairo, and British Embassy documents were destroyed by fire lest they should fall into enemy hands — the day became known, because of this act, as "Ash Wednesday". Many Egyptians openly mocked, insulted and spat upon British and French officers and soldiers in Cairo. In "Alex" the Navy withdrew its fleet and its shore personnel, and the city was stripped of its defences and declared "open".

Everyone however had reckoned without Auchinleck — except perhaps Rommel, who always respected the British and expected counter-strokes. Having assumed direct command of his troops the "Auk" fought and won the first Battle of El Alamein and stopped the Germans in their tracks. Later in the month the battles at Tel-el-Eisa and Ruweisat Ridge forced the Desert Fox to dig in and lick his wounds.

Churchill came out to Egypt to visit Auchinleck on August 7th and decided to remove him. In his place as C-in-C, Middle East, he appointed General Sir Harold Alexander, another Irishman. For 8th Army, against the advice of the Chief of the Imperial General Staff, Sir Alan Brooke, he appointed as GOC Lt General "Strafer" Gott. However Gott was killed in an aircraft shot down by an enemy fighter before he could assume his command and Brooke's choice was then appointed — Lt General Bernard Law Montgomery who took over command of 8th Army on August 13th.

By the time "Monty" arrived 9th Regiment had moved out of Reserve to join not 8th Army, but 9th Army in Palestine. They left Qasassin on August 1st, crossed the Suez Canal and then travelled through the Sinai Desert to Jerusalem which they reached on the 6th. Regimental HQ was established at Ramleh, 24 Battery moved to Haifa under the command of 26 LAA Regiment, one Troop of 25 Battery went to Ramleh and one to Aqir while 26 Battery was split between Lydda and Sarafand. 115 LAA Bty came under the Regiment's operational command in the defence of strategic bomber fields of both the RAF and the US Army Air Force. Two weeks later a Troop of 26 Battery plus BHQ moved to Rayak in Syria (now in Lebanon) leaving the remaining Troop attached to 25 Battery.

In early October 24 Battery moved back under the command of the Regiment occupying sites at Aqir and Sarafand. Since early August they had been based at Ramat David and St. Jean with BHQ at Caesar's Camp, Acre. They had had one "stand-to" on September 3rd when enemy bombers had attacked the defended area. As friendly fighters were in operation however the guns did not fire. 242 HAA Battery arrived to relieve them on October 6th and the following day 24 returned to their parent unit.

BHQ and Right Troop, 26 Battery had been ordered to move to defend Rayak aerodrome in Syria on August 23rd. They arrived on August 25th and began to prepare positions for the guns and instruments and to set up a small GOR. A troop of LAA guns was also assigned to them. They saw no action at Rayak before their relief by 279 HAA Battery and they returned to 9th Regiment on October 13th, taking over sites at Lydda. While at Rayak one of their number, Gnr James McNeill had collapsed at his gunsite and died while being taken to a Casualty Clearing Station. Gunner McNeill's death was later discovered to be due to "toxaemia, cause not known".

25 Battery had remained at its sites at Ramleh and Aqir, where they had also had to prepare and build positions. There was no action in this area but a scheme of mobile training started on August 31st.

With the entire Regiment together again by mid-October Regimental mobile training schemes commenced. Activity was hampered by very heavy rain in October and they endured one series of thunderstorms which lasted for 18 hours. Gunsites and accommodation huts were flooded and some guns had to be removed to ordnance for overhaul after being submerged. On October 17th two gun tractors of 26 Battery which had been left in a dried-up river bed were marooned by rising flood-water. There were soldiers on both lorries, and after two attempts L/Sgt J B ("Tich") Tichbourne managed to swim out to the vehicles with a rope to lessen the danger to the marooned crews.

During November the Regiment carried out an extensive series of mobile exercises during which each Battery would deploy to defend positions in the Jerusalem and Dead Sea areas. These exercises included air co-operation and anti-tank gunnery as well as negotiating the difficult Seven Sisters Road — a piece of experience that was to prove useful later at the Derna Pass. Each exercise lasted 48 hours and all three batteries participated, although 24 Battery had returned to Ramat David and St. Jean for two weeks in early November.

Their stay in Palestine was nearly over and although there had been no action it had been a memorable time. Many had taken the opportunity to visit the holy places of Jerusalem as well as Bethlehem and Nazareth and the Regimental Chaplain, the Rev Louis Crooks, had in fact organised sight-seeing tours.

Strabane man Johnny McMahon had even written home breaking all the rules about not divulging their location by telling his readers that "I am now in the land where Christ was born but I wish to C....t I was back in the land where I was born". No doubt the censoring officer was so amused that he let the letter go through without obliterating the offending statement.

The warning order to leave Palestine was received on November 22nd. Three days later RHQ was told that the Regiment was to join 8th Army. On November 28th they left Palestine and on December 1st were once again at Qasassin.

CHAPTER THREE

THE TIDE TURNS — EL ALAMEIN

When Auchinleck brought Rommel's forces grinding to a halt at El Alamein on July 1st 1942 it was little appreciated then that the tide of German advance had been halted. The early months of the year had seen the high point of Axis success across the globe, but Midway and First Alamein now checked them.

Auchinleck's replacement by Alexander and Montgomery brought little immediate change to the men on the ground, who were now not disposed to trust any officer outside their own company, battery or squadron. The little man who had taken over 8th Army was an unknown quantity to them, even though he had had a distinguished career in the Great War and had also checked Rommel in the German attack on France in 1940. Within months however Montgomery was to become a household name as he led 8th Army to victory in the autumn of 1942. Their first reaction however was one of cynicism, but this began to change when Monty accurately predicted that Rommel would attack within a month adding that, for 8th Army, there would be "no further withdrawals". Not only did he predict accurately the timing of Rommel's attack but also where it would happen — an attempt to turn the southern flank of the Alamein line. Such a manoeuvre had worked for the Desert Fox at Bir Hakeim in May but at the end of August there was another fox in the desert. (The truth of Monty's "prediction" however was that British Intelligence had been able to decipher German signals since the beginning of the war but this fact was only made public in the mid-70s.)

The second battle at El Alamein, the Battle of Alam Halfa, began on August 30th. By September 2nd Rommel was forced to withdraw his attacking units. It was his last offensive in the Western Desert.

For the next six weeks or so 8th Army, now consisting of three corps (XIII, XXX and X) built up its strength and prepared for the offensive which Monty promised would push the enemy out of North Africa. Now his soldiers believed him!

Rommel, ever the aggressive, had been forced to dig in to prepare to meet an attack. His defences were deployed as had been those of the Western Front a generation before, with over a million mines laid along with miles of barbed wire; Montgomery, who valued dearly the lives of his troops, would be forced to fight a Great War battle.

The plans for the Alamein offensive included an elaborate and, as it transpired, successful deception plan which led Rommel's staff[1] to miscalculate the point at which their adversary would attack.

Operation "Lightfoot" began after 9.00 p m on October 23rd 1942 when RAF Wellington bombers dropped over 120 tons of bombs on the rear lines of *Panzerarmee Afrika*. At 9.40 p m almost 900 field and medium guns opened fire in the greatest artillery barrage since the First World War. The guns fired for fifteen minutes on the known positions of enemy artillery batteries pounding them into scrap metal. So intense was the bombardment that it could be clearly heard 60 miles away by the Franciscan Sisters in Alexandria's Sacred Heart Convent! The infantry attacked at 10.00 p m moving through the minefields in moonlight, the pipes of the Highland Division adding their skirl to the noise of battle.

The Alamein battle continued for days, with Montgomery revising his tactics through Operation "Supercharge" on October 30th. By November 4th Rommel had

[1] When Montgomery launched the Battle of El Alamein Rommel was not in command of the Axis forces having been recalled to Germany to recuperate from illness. Thus the initial attacks were met by Axis troops under General Georg Stumme who died of a heart attack shortly after the Battle began. His place was taken by the commander of the *Afrika Korps*, General Ritter von Thoma until Rommel returned to Egypt.

decided to withdraw and in a brilliant manoeuvre, so typical of the man, he pulled the remnants of *Panzerarmee Afrika* out of 8th Army's clutches. Behind him he left 30,000 prisoners, including the entire Ariete Division, with most of his tanks and artillery shattered and thousands of dead and wounded. 8th Army had sustained some 13,500 dead and wounded in the twelve days of battle. Monty had predicted accurately the duration of the battle and its casualty toll. It had not gone entirely to plan — thus the regrouping for "Supercharge" but in the end a decisive victory had been gained and the pursuit of the fleeing Rommel was underway.

8th Army now raced to Fuka in an attempt to cut off and capture the remaining enemy forces but Rommel again slipped through. Then the weather came to Rommel's aid, heavy rain slowing down Montgomery's men at Mersa Matruh, but by the end of November Montgomery had captured Benghazi — it was to be the last time the city changed hands.

With the front line now so far west the airfields of Palestine were no longer of such great importance and thus 9th Regiment was withdrawn to Egypt to prepare to move up to join 8th Army.

The Regimental convoy left Qasassin on December 3rd at 6.15 a m and travelled westwards with nightly halts at Wadi Natrun, El Alamein, Mersa Matruh, Buq Buq, Capuzzo, Tobruk, Tmimi and finally Martuba which they reached on December 11th. Heavy rain caused problems near Capuzzo forcing the convoy to retrack from the Trigh Capuzzo and to use the main coast road.

26 Battery were detached from the Regiment at Tobruk to provide air defence for the port. On December 12th they were in action against a formation of enemy aircraft. Left Troop fired 75 rounds on radar control and Right Troop 12 rounds of barrage fire. Two aircraft were destroyed and credited to the joint efforts of the RAF and the gun defences.

Next day Right Troop/26 Battery moved to defend the railway sidings at Capuzzo. Due to a derailment there was an excessive build-up of trucks in those sidings providing an inviting target for aircraft. A recce plane had already been over and an attack was feared. Right Troop returned to Tobruk on December 18th after the sidings had been cleared. That evening they were again engaged, together with Left Troop, in an attack on Tobruk during which 53 rounds were fired and an aircraft was brought down. This aircraft was also jointly credited to the RAF and the guns. Further raids occurred on December 19th when 103 rounds were fired and on Christmas Eve when a single Ju 88 divebombed the harbour and Right Troop fired 11 rounds.

On December 31st the Battery was relieved at Tobruk and moved to Martuba to rejoin 9th Regiment. At Martuba and Derna 24 and 25 together with 117 LAA Battery had deployed in the defence of airfields, 25 going to Derna from which it was intended to launch a heavy bomber strike against Crete. However the weather during this period had been so bad that it had severely restricted both Allied and Axis air activity, and the guns were not involved in any actions.

Montgomery had galvanised 8th Army into believing in itself. He was, as one writer expresses it, the "dynamo" of that army. Now each man believed that his destiny and that of his army was victory. Churchill summarised their feelings and attitude when he said that "before Alamein we never had a victory; after Alamein we never had a defeat"[2]. Appreciating the strategic significance of the victory the Prime Minister had ordered the church bells of the country, which had been silent for three years, to ring out in tribute.

CHAPTER FOUR

TRIPOLI FALLS

An order to move up to Benghazi was received by 9th Regiment on January 3rd 1943 and they accordingly arrived there the next day. They remained near Benghazi for two days and those in fact were the only two days of the War when the Regiment, as originally constituted, was together. 6 LAA Battery had been in Benghazi since November and this was the first occasion on which they met up with their "parent" unit.

9th Regiment was now ordered to take part in the advance on Tripoli and therefore left Benghazi to travel into Tripolitania, through Mussolini's triumphal arch — "Marble Arch" — to join up with 51st Highland Division, west of Nufilia on January 11th. The Regiment was constantly on the move close behind the front-line troops for the next twelve days. 25 Battery, which Col. Aitken considered had performed best in the mobile exercises in Palestine, was placed under the control of the Commander, Royal Artillery (CRA) of 51st Highland Division and deployed on a mobile basis to protect the advance of the "Jocks". 26 Battery was engaged in protecting troop concentrations and then in covering Royal Engineers engaged in mine-clearing operations.

Only the main road and a track, known as Highland Division track, were cleared of mines and the advancing army had to take great care. Even so there were many casualties from the mines but fortunately not within 9th Regiment. There were occasions when there were close shaves however. One of these occurred when a gun and tractor of 25 Battery got into difficulty. One of the gunners jumped over the tapes, laid by the sappers to mark the dangerous areas, to assist in the rescue operation and he returned the same way. Moments after the lorry and gun moved off a mine exploded just where they had been stopped! As they got closer to Tripoli the road became more congested and it was impossible to leave it due to mining even when "laagering" at night.

[2] Churchill obviously either forgot or overlooked the fact that there had been victories before Alamein and indeed two of those had been in the Desert — Wavell's victory in 1940 (Operation Compass) and Auchinleck's in 1941 (Operation Crusader). There had also been Cunningham's successful campaign in Abyssinia. Nonetheless Alamein was a turning point and Churchill presumably granted himself a certain degree of licence in order to emphasize this point.

On the road to Tripoli. The greatcoats indicate that North Africa was not always warm and sunny. Standing centre is J J Doherty in Despatch Rider's helmet and carrying a personal sidearm. Standing right is J J Ormsby.

(Photo: John J Ormsby)

On January 20th the entire Regiment came under command of 51st Division and they moved on to Highland Division Track as the main road was now completely blocked. Their orders at this time were to be ready to establish anti-aircraft defences in Tripoli as soon as the city fell and so they were given priority on the advance with elements of the Regiment being among the first British troops to enter Tripoli on January 23rd.

Next day the entire Regiment had deployed for the defence of Tripoli. Lt. Col. Aitken was appointed AA Defence Commander (AADC) of Tripoli and several other units passed under his command. These were 94th HAA Regiment, 53rd and 40th LAA Regiments and 306 Searchlight Battery. All of 9th Regiment's guns were in position ready for action at 4.30 p m that day.

The defence of Tripoli harbour was to be the Regiment's task for almost six months and it was to be a task that it would perform superbly under tremendous pressure from the enemy. Tripoli was vital to Montgomery's plans as he advanced towards Tunisia. His supply lines had been stretched all the way through the desert: Tobruk and then Benghazi had helped to shorten them. Now Tripoli, the biggest harbour in North Africa after Alexandria, would enable the re-supply of 8th Army to be carried out much more efficiently.

First however the harbour had to be cleared. Subjected to heavy bombing by the RAF and demolition work by the retreating Germans and Italians it could not yet accept ships. The work of clearance was undertaken by the Royal Engineers with the help of working parties from the various regiments in the Tripoli area. By the beginning of February the harbour was capable of taking the first ships, although small, and when the Prime Minister visited Tripoli on February 3rd and 4th he was able to watch ships being unloaded. To the labouring soldiers, including men of 9th Regiment, Churchill said "You are unloading history".

The PM's visit was also marked by a victory parade in Tripoli, the first enemy capital to fall to the Allies who, since the beginning of the year, had adopted the title "United Nations". In this parade, reviewed by Churchill and Montgomery, each of the Batteries of the "Ninth" provided a party to march with the 51st Highland Division. Those who were close enough to Churchill could see that there were tears in his eyes as he took the salute. Earlier he had addressed men of the 8th Army, including many of the "Derry Boys" and told them ". . . that after the war when a man is asked what he did it will be quite sufficient for him to say, 'I marched and fought with the Desert Army'. And when history is written and all the facts are known, your feats will gleam and glow and will be a source of song and story long after we who are gathered here have passed away".

Already the *Luftwaffe* had turned its attention to the harbour. On January 25th a Ju 88 had carried out a reconnaissance over Tripoli flying from south-west to north-east. Engaged by one gun site only which fired 9 rounds the plane made off without damage. Two days later a fighter-bomber dropped one bomb west of the town during the night. It was not engaged, but a minelayer which came over in the morning had 182 heavy rounds and 560 light rounds fired at it.

The month of February was to see a determined onslaught on Tripoli with regular attempts to bomb the harbour or to drop mines in the vicinity. For the defenders there would be little rest as they brought their guns into action no less than 29 times. There were occasional days on which no raids would occur but, as if to balance this, on other days up to three raids could be launched, some lasting as long as three hours.

The scene was set for the February battles when on February 2nd a low-level attack took place, carried out by two aircraft at 7.05 a m. All three batteries fired a total of 120 rounds. Just after noon a Ju 88 flew over on a recce flight; it was engaged by 26 Battery only with 5 rounds. They described their fire as "poor" however, although the Junkers was flying at 23,000 feet, which did not make it the easiest of targets. Then at 5.30 p m at least six aircraft took part in a bombing raid on the harbour. Shooting was difficult since dusk was falling, and so barrage procedure was used. The raid lasted for an hour during which the aircraft attacked singly. Bombs were dropped but no damage was done as the barrage of over 1,000 heavy rounds deterred the attackers from pressing home their attacks too enthusiastically.

On the first day of Churchill's visit, February 3rd, the *Luftwaffe* stayed away but on the second day when the Victory Parade was held a recce plane came over. It was chased off by the anti-aircraft guns, as was a single intruder which approached the harbour that evening.

There were three raids on February 5th, the first of these being a bombing attack at dawn. At least seven bombers were involved, only one of which ventured close enough to drop a stick of bombs north of the shipping outside the harbour. One bomb did fall in the city however, causing some civilian deaths, and a number of mines also fell in the harbour approaches. The fire of 9th Regiment's batteries totalled over 800 rounds and the gunners were becoming attuned to the high firing rates that were going to be demanded of them later in the month. Less than three hours later a recce plane was engaged by 25 and 26 Batteries and that night harbour barrages and GL directed fire was used during a minelaying raid by about four aircraft.

A Messerchmitt 210 came over on a recce flight just before 2.00 p m next day to be met by a salvo from 24 and 26 Batteries.

Recce aircraft had been appearing on a daily basis as the *Luftwaffe* tried to build up an accurate intelligence picture of the use of Tripoli harbour for its *Kampfgeschwader,* or bomber groups, based in Sicily, to whom the task of denying the use of Tripoli to Allied shipping had been given. In spite of this obvious military purpose for these flights many of the gunners concluded that there was in fact an element of personal vindictiveness in them. This was especially true for the men of 24 and 26 Batteries who had had similar regular experience in Alexandria and had long since decided that the midday recce was really "the so-and-so who was sent over to make sure we got no rest at all"! There was to be very little opportunity for rest for the gun crews around Tripoli Harbour but unlike other locations few attempts were to be made to hit the guns themselves.

February 7th saw a fine example of the degree of attention which the Germans were preparing to give the harbour. At 7.35 p m at least three bombers made a sharp raid on the defended area. The searchlights illuminated two machines which were vigorously engaged by the guns. 9th Regiment's guns fired almost 700 rounds over a forty minute period and claimed to have damaged two aircraft. However Regimental HQ only recorded one of these as having definitely been damaged.

Next morning another "damaged" was logged, this being credited to 153 HAA Battery. The aircraft, a Heinkel He111 on a recce mission, was later attacked by RAF fighters and claimed as a "probable" by them. That evening the harbour again came under attack from a group of at least seven aircraft. As the guns hammered away the track of one plane disappeared completely from the radar screens. Its destruction was claimed by both 24 and 25 Batteries but the available evidence tends to support the 24 Battery claim as A Troop/24 Battery had been firing at an aircraft caught by searchlights which had been hit by gunfire and was then seen to fall rapidly. Regimental HQ recorded the aircraft as a "probable"; two other attackers were brought down by the RAF's fighters. 24 Battery felt quite pleased with their evening's work as their guns had put up more rounds than the other two batteries combined.

After this hectic spell there was a brief respite until shortly before dawn two days later when several aircraft approached Tripoli. No attack developed on the harbour, as the attackers were turned away by a short fierce barrage of more than 200 heavy rounds accompanied by over 400 light rounds. There was a change in the LAA regiments in Tripoli that day with 14th (New Zealand) LAA Regiment replacing 40th LAA Regiment. The heavy defences had been strengthened by an additional regiment, 51st HAA Regiment, which had recently come under the command of Lt Col Aitken as AADC, Tripoli.

The "usual" recce plane was seen later that day but it was not until 7.00 a m on February 14th that another attack occurred when a single mine-laying bomber was engaged by barrage fire after being illuminated by searchlights. Several shells were seen to burst close to the plane which lost height rapidly and disappeared from the radar screens. The Regiment claimed this as a "probable". Barrage fire turned away attackers that evening and no bombs were dropped near the harbour. 9th Regiment's guns had really excelled themselves as they had put up almost half of the 687 heavy rounds fired in the engagement.

Three single aircraft, including a recce plane, were engaged by 24 Battery the following day. The recce aircraft turned away while the guns ceased firing at one of the other attackers as RAF night-fighters closed in. Then followed the longest spell without raids during February although 25 Battery did fire three rounds as indicators to guide fighters on to a recce plane on the 16th. Not until the morning of the 19th however did the guns fire again in defence of the harbour when six or more aircraft on a bombing and mine-laying raid were engaged by harbour barrage and GL directed fire. Some bombs did fall on the harbour front, one of them damaging three vehicles of D Troop, 14th (NZ) LAA Regiment. There were no casualties however.

General Montgomery visited the Regiment on February 20th. While he was in B Troop/24 Battery's position A Troop/24 and B Troop/25 opened fire on a single plane. As this occurred around midday it was probably the recce flight which was still regarded as a real nuisance by the gunners. Monty spent some time chatting to the men and was especially pleased to meet men from Londonderry in which city his

General Montgomery with members of 24 HAA Battery at Tripoli. In the foreground is a Height and Rangefinder.

(Photo: Tommy McCready)

family had its roots — his family seat was then at Moville less than 20 miles from the city. Tommy McCready recalls Monty talking to him and asking "What part of Derry do you come from?". The overall impression of their GOC in the Regiment was of a man who cared for the well-being and morale of the individual soldier. This was reinforced when, after an official photography session, Monty told the men that they could go and fetch their own cameras if they wanted group photos with him. "Not a man moved", remembers John Ormsby, "for although we had cameras we knew that it was an offence to have a camera in a war zone. Then Monty told us a second time to get our cameras and when nobody moved yet again he realised what was wrong. 'I know you have cameras' he told us, 'so do fetch them. That is an order from your General!' ". Needless to say it was an order which was obeyed and showed Monty's instinctive ability to communicate with the man on the ground.

69th HAA Regiment moved into Tripoli on February 21st to relieve 94th HAA Regiment and the newcomers were soon in action as a formation of at least eight bombers attacked the area that evening. Once again no damage was caused to the harbour and the massive barrage put up by the heavies (almost 2,500 rounds) not only blunted the ferocity of the attack but accounted for one of the attacking aircraft.

Before morning broke on the 22nd over twenty aircraft swept in on a bombing raid and were met by concentrated fire in the form of barrages and GL and height-control fire. The attackers took a heavy punishment as the guns blazed away for over thirty minutes. In all, the three HAA regiments put up over 3,000 rounds of which total 9th's batteries fired more than a third. Three German planes were seen to crash and were credited to the guns, although 25 Battery were certain that the AA guns had brought down five attackers. Two others were destroyed by the RAF and another two or three were damaged by the fire of the ack-ack gunners. Of those which came down A Troop/24 Battery claimed the destruction of one. Regimental HQ later confirmed that three aircraft were brought down by AA fire — two by HAA and one by LAA — and noted that a fourth had probably been destroyed by the heavy guns. There were reports that another two crashed into the sea after being hit by AA fire. As many as seven or eight aircraft lost was a heavy price for the *Luftwaffe* to pay for no return as the raid had inflicted no damage at all!

The next raid, which came before midnight on the 24th, was turned by barrage fire, and again there was no damage to the harbour. In the early hours of February 26th the *Luftwaffe* returned with a force of six bombers to attack the harbour. Once again an intense barrage was put up with 4,362 heavy rounds being fired in an hour! About one third of these came from 9th Regiment's batteries, who claimed one bomber destroyed as it fell in flames to the west while another was seen to dive steeply towards the sea but was not seen to strike the water. One "destroyed" was confirmed.

Another heavy raid followed twenty-four hours later with eight bombing and mine-laying aircraft being met by yet another intense barrage. Once again the attackers had no success and turned for home, this time without any definite losses for no hits were claimed by the defenders. 25 Battery's contribution had been less than normal as B Troop had left their guns at a practice-range at Azizia where they had been carrying out field-gunnery training with the 3·7s.

The *Luftwaffe* saw February out in style with three attacks on Tripoli on the last day of the month. For almost two hours from 3.23 a m intruders dropped bombs and mines and were countered by barrage fire, almost 1,000 being fired by the "Derry Boys". This raid occurred in two phases, the first being a "half-hearted effort" between 3.23 and 4.00 a m. Then the "usual" raid began with at least eight bombers coming in at intervals from the north and the north-west. Their bombs fell well clear of the harbour but one very large flash seen at sea-level could well have been an aircraft hitting the sea. More determination was shown in the next attack just before 7.00 p m. Indeed one aircraft flew straight across the harbour in spite of the barrage. This was so unusual as to be worthy of note in the War Diary. It certainly must have been a frightening experience for the crew of the aircraft! Yet again the attackers were met with determined opposition and although no claims were made the bombers once more made off with little effect. At 10.20 p m the third raid of the day began. RAF nightfighters were in the air and shot down one mine-laying bomber, while the heavy guns put up a relatively light total of under 200 rounds.

Thus since the capture of Tripoli its defenders had successfully fought off 34 raids, comprising 23 bombing or mine-laying attacks and 11 reconnaissance aircraft. At least five, and possibly as many as ten, enemy aircraft had fallen to the heavy guns, another to the light guns and three to fighters with up to seven "probables" and damaged. 9th (Londonderry) HAA Regiment, still the commanding unit, was the only regiment which had been continuously in service in Tripoli since its capture and was now grouped with:

51st HAA Regiment
69th HAA Regiment
30th LAA Regiment
57th LAA Regiment
306th SL Battery

March began with the *Luftwaffe* airmen already showing a healthy respect for the Tripoli defences. By the end of the month that respect would be even greater as the Germans bestowed the title *"Wand aus Stahl"* (Wall of Steel) on the Tripoli harbour AA defences.

For the men on the ground it had been an exhausting time yet also, in a strange way, an exhilarating time. They had been so busy that there had been no time for boredom and, in spite of the continuing danger, there was that feeling that the war had turned a crucial corner. Morale was high and each gun-crew was determined to show that it could fire more rounds than its neighbours, with a similar rivalry between Troops and Batteries.

The firing rates achieved on some of the February raids were to be equalled and sometimes exceeded during March. Obviously such high rates of fire put a great strain on the ammunition supplies and on the personnel whose job it was to keep the guns supplied. Robert Hamilton recalls one occasion when the ammunition for his gun was about to run out — and still the bombers were coming in! Exasperated by this he called to the men who were bringing the rounds from the ordnance dumps to the guns, "Get more —— ammo up here and make it fast!" only to hear himself being reprimanded by Captain Paddy Laird from the Command Post — "there's no need for that language, Sergeant Hamilton, your ammunition is on its way"!

Heavy raids brought almost everyone in the Regiment into action in one way or another including the cooks. Jim Canning recalls that the cookhouse used to be ordered to boil up large quantities of water so as the barrels of the guns could be sluiced out both to clean them and to cool them down during sustained periods of action. Cold water could not of course be used on metal that was already glowing cherry-red for fear of damaging the barrel liners.

Although the Germans were not having a great degree of success with their bombs there was still plenty of metal falling on the gun positions as the remains of their own shells fell to earth. Some of the splinters could still be red-hot and even without the heat the danger caused by them meant that steel helmets had to be worn all the time when in action and the discomfort suffered by a perspiring gunner in a "tin bowler" simply had to be endured.

The noise was horrendous as over seventy guns crashed forth in barrage fire. At one of 25 Battery's positions, in an old Italian gunsite with accommodation dugouts beside and below the gun, a fellow Derryman who was serving in an infantry regiment was a visitor when a raid began. He was told to stay "down below", but found the experience so terrifying as the guns shook the earth and their thunder blasted his senses that he emerged into the gunpit and offered his services as an ammunition bearer! (Nor was he the only man to find such an experience to be something that he did not want to repeat; Colonel Jack Christie tells a similar story about Blair Mayne who visited his Battery HQ (6 LAA Bty) at about this time and left after experiencing an air raid. Mayne had intended to stay two nights with Major (as he then was) Christie but he reckoned that life in an AA unit was "too dangerous"!! Mayne had of course been commissioned into 5 LAA Battery!)

March began however with the Regiment providing 36 men to join with personnel from 51st HAA Regiment to form a special troop manning four captured German 88 mm guns which were to be used for counter-battery work. The Troop was commanded by Captain Downing of 51st HAA Regiment with Lieutenant Stewart-Moore of 26 HAA Battery as second-in-command. In the tradition of naming such ad-hoc units after the person responsible for initiating them the 88 mm Troop was called MAC Troop after the commander of 2nd AA Brigade, Brigadier H M J (Mad Mac) McIntyre[1]. MAC Troop was to serve with New Zealand units in the field, and in fact its inception was due to the impression made on the CRA of the New Zealand Division, Brigadier Weir[2], when he visited Azizia and saw 25 HAA Battery demonstrate its 3·7s in a field role. 3·7s had been used as field guns before, during the siege of Tobruk in 1941, but that had been considered as a matter of expediency and seemingly was not regarded as a precedent. The 88s were regarded as multi-purpose weapons,

[1] Brigadier McIntyre's nickname is believed to have arisen from his habit of joining in AA Barrages with his service revolver. Needless to say he was an exceptionally popular officer.

[2] Weir was accompanied by General Sir Bernard Freyberg VC, the GOC of New Zealand forces in North Africa.

much more so than the 3·7s, but the latter would distinguish itself as a field piece in Italy. (The story of MAC Troop is included in Appendix 3.)

The first raid in March took place on the 2nd and lasted for forty-five minutes from 6.25 a m. About ten aircraft were involved and they were very heavily engaged. One machine was illuminated by searchlights and shot down while a "probable" was also claimed. Over 5,000 heavy rounds were expended, yet 25 HAA Battery's War Diary does not mention this raid! Nonetheless 25 Battery men are adamant that their Battery was involved and it seems most unlikely that it would not have been. It can be reasonably assumed that the Battery's fire equalled that of each of its fellow batteries of 9th Regiment. Regimental HQ noted that the intruders tried to confuse the defences by circling the area and then pushing home sudden attacks "with determination". The rate of fire maintained by the heavies was also described as being "excellent" and indeed it averaged 75 rounds per gun! The cooks would certainly have been busy with boiling water that evening. One of the searchlights (No 9) was destroyed when a bomb fell on the Spanish Mole but fortunately there was no loss of life.

Five days passed before the bombers came again. There was the feeling that the *Luftwaffe* was trying to lull the defenders into lowering their guard, but the readiness levels were not allowed to drop. Because of this the guns were ready when about fifteen aircraft attacked in waves from 6.45 p m to 7.09 p m on March 7th. In spite of the tactics used by the bombers they had little success even though they dropped about 40 bombs. Two small ships in the harbour suffered some minor damage but one bomber was destroyed (claimed by 24 Battery) and another was classed as a "probable". At the same time next evening two intruders turned away in the face of a harbour barrage when they were still some 15,000 yards from the harbour itself.

On March 8th the Regiment ceased to be part of 8th Army as that formation moved further away from Tripoli, but the value to 8th Army's operations of 9th Regiment did not diminish in any way and of course the Regiment was still represented in 8th Army through MAC Troop. Other administrative changes which took place at this time included the relief of 57th LAA Regiment by 34th LAA Regiment. On March 11th 34th LAA Regiment sent a battery to Castel Benito airfield, to which location a heavy battery had also been moved from Tripoli.

The evening of March 11th saw another attack on the harbour but it was not pressed home with any real determination, although some incendiaries fell near a hospital. No hits were claimed by the AA defences which fired off almost 2,000 heavy rounds. A low-level attack without any early warning occurred on March 15th when two aircraft dropped some bombs on the harbour, but no damage was done and the attackers made off before the guns could put up more than 44 rounds. Two hours later however more than a thousand heavy rounds were fired at a further trio of attackers.

One of the heaviest barrages at Tripoli was put up the following evening. Thanks to the excellent work of the searchlights which lit up and held five of the attacking planes four enemy aircraft were shot down by the AA guns. Just over 5,500 heavy rounds had been fired by the Tripoli defenders and 9th's guns had contributed more

A 3·7 of 26 HAA Battery at Tripoli.

(Photo: Tom Moore)

than their share. Once again the firing rate had been very high; 24 Battery had averaged almost 31 rounds per minute for each of the 25 minutes during which the Battery had been in action and there were similar pictures at 25 and 26 Batteries.

On the evening of March 19th the *Luftwaffe* almost succeeded in obtaining total surprise for a raid. There was heavy rain and low cloud and out of this suddenly appeared a formation of about fifteen bombers. No radar warning had been received, for obviously the aircraft had flown low and therefore "under the radar", while the weather conditions had muffled the sound of their engines. But the gunners were on "stand-to" in spite of the conditions, and the guns quickly opened fire. The first gun to fire that evening was Sergeant Robert Hamilton's at T 4 site near the harbour entrance. To this day Robert Hamilton is convinced that his first round still had its safety cap on, as a count next day of the "capped" rounds on the site showed one short! If it did strike a German plane the round would have passed straight through without exploding but could nevertheless have inflicted considerable damage. At first the guns at T 4 fired independently on "gun control" until the entire defences had been alerted and co-ordinated barrages could be put up. Several of the planes flew right into the barrage fire and exploded in mid-air. The guns brought down six aircraft, and there were one "probable" and one damaged. In a period of just twelve minutes over 1,300 heavy rounds had been fired. Small wonder therefore that the events of that evening are still very clear in the minds of those who were there. But for the first time ever 9th Regiment saw ships under its protection hit. One merchant ship was struck by bombs and sank; another, containing ammunition, was hit by one of the crashing bombers and blew up. And a destroyer was damaged and had to be beached. This however was the *Luftwaffe's* only success at Tripoli and it had been bought at a terrible price.

No air force could continue to sustain the losses which the *Luftwaffe* was suffering over Tripoli. The effects on morale of these seasoned aircrew — the *"Experten"* — were enough to make even the bravest man more cautious. And so, not surprisingly, this was their last big raid on Tripoli.

The guns had won the Battle of Tripoli Harbour and for his services there Bob Aitken received the OBE, an award which he considered to be as much that of the men in the gunpits as it was his. Although there would be further raids on the harbour they would be minor compared to what had gone before. Throughout, until the last raid, the constant factor in the defence would be the guns of 9th (Londonderry) HAA Regiment.

CHAPTER FIVE

TRIPOLI HARBOUR DAYS — AND NIGHTS

Those days at Tripoli in 1943 were without doubt the apotheosis of the Regiment's time in the AA role. The defence described in the previous chapter drew comment not only from the *Luftwaffe* who had to face it but also from RAF airmen flying in defence of Tripoli. In fact the RAF described the barrage as the most fearsome in the Middle East and one RAF officer quoted in the "Tripoli Times" said — "The most unpopular order to receive in the German air force at the moment must be to set out on a raid over Tripoli". Some of the RAF night-fighter crews had operated over Britain during the blitz but like their *Luftwaffe* foes they were in no doubt as to where the fiercest AA barrage was to be found. Tripoli was the "barrage of barrages" and it was said that "no plane can live in that harbour barrage. The sky is full of flying steel".

The truth of these statements is shown by the toll of the barrage on the attackers and the insignificant effect that the German attacks had on Tripoli harbour. 9th (Londonderry) Regiment was the only AA unit involved in the Tripoli defence from start to finish. In fact their presence as an essential part of the defence scheme was due to a request from the Royal Navy before the city even fell to the Eighth Army; mindful of the excellent work done by the "Derry Boys" at Alexandria the navy wanted to be sure that the Regiment, which was the most experienced AA unit in the Middle East, would be involved in the defence of the vital port of Tripoli. The faith of the navy in the Regiment was not misplaced, for they more than proved their worth in the weeks that followed. Although there were two other heavy regiments involved at Tripoli for much of the time an analysis of the rounds fired shows that Ninth did more than their share of the firing.

There is no doubt that their spirits were high in those days as they sweated and strained for hours on end to build their "Wall of Steel". At times their firing rates were so high and the action so prolonged that the barrels of their 3·7s glowed red and boiling water had to be poured down the barrels to cool them! (see previous chapter.)

The Tripoli "barrage" was in reality a series of barrages, any of which could be used depending on the nature of a particular attack. The HARBOUR BARRAGE was a "box" around the harbour area with some guns firing at 5,000 feet and some at 8,000 feet. It was fired in sequences of four rounds per gun. The first round was fired on the bearing ordered by the Ops Room; the next was 4° to right of that bearing; the third was on the original bearing and the fourth 4° to left before the sequence was started again. The firing rate was "the maximum possible". A BLACK BARRAGE was a semi-circle of 2,000 yards radius to seaward of the outside anchorage and was designed to protect shipping in that anchorage against low-flying bombing or torpedo attacks while a BLUE BARRAGE was a similar tactic designed to protect ships inside the harbour. There were other "barrages" such as SUN and SPIDER barrages included in the defence programme which was devised by Lt Col Aitken whose orders were regarded as being refreshingly simple in comparison to the lengthy missives in other areas. And yet that simplicity proved so effective.

Obviously there was little time for rest. Guncrews were at readiness on a 24-hour basis and in the early evening, the most likely time for attack, all gunners had to "stand-to" the guns for instant reaction. On that evening of March 19th when the bombers came charging through the cloud and rain it was this high state of readiness which denied the *Luftwaffe* the element of surprise for which they had hoped.

For those not on gun crews the days were kept busy with helping to unload supply ships and with transporting stores from the harbour to the various supply dumps. Somehow or other Tripoli became a "home from home" for 26 Bty whose veterans remember their Tripoli days with affection, for living conditions had improved somewhat, the Regiment was together as a unit and there was the distinct feeling that the tide of the war had really turned for the Allies. As if by way of celebration a large number of 26 Bty NCOs and gunners promoted themselves to officer status — at least in terms of footwear as they became the proud possessors of a quantity of "misplaced" brown shoes!

CHAPTER SIX

PREPARING AN AVALANCHE

Of course there was no way in which the defenders of Tripoli could know that they had had their last major raid. The defences continued to be maintained at a high level and indeed at the end of March were strengthened further. Fierce fighting was still raging in Tunisia, where the Battle of the Mareth Line was about to take place. Although Rommel had left Africa on March 6th, a sick man, his place had been taken by General von Arnim who was to prove an equally tough opponent for 8th and 1st Armies.

A number of recce aircraft were fired on in the closing weeks of March but no serious raids happened. On March 29th an additional four heavy guns were added to the defences; these were captured Italian 90 mm guns which were manned by a composite troop from 9th and 51st Regiments. Lieutenant R R Sands of 24 Battery was appointed Troop Commander and Tommy McCready was one of the 24 Battery men attached to the Troop.

On March 31st a further administrative change took place and Troop Commanders were promoted to the rank of Captain. Each Battery now had three Captains — the Battery Captain (second-in-command) and the two troop commanders. Within the Batteries the Sections remained as sub-units, each commanded by a Lieutenant; their designations had changed some time before and were now numbered throughout the Regiment, i e 1-4 Sections in 24, 5-8 Sections in 25 and 9-12 Sections in 26.

April began with a visit by Major-General Pollock, the AA Group Commander, Eastern Mediterranean. Pollock visited the GOR and a number of the gun positions.

Recce aircraft flew over on April 2nd, 4th and 5th. The first machine, a Ju 88, was chased across the harbour by a Spitfire while the others were fired at by the heavy guns.

With the Regiment adopting a semi-mobile role it had a number of surplus personnel and these were posted out, some of them going to 6 LAA Battery which was now at Misurata near Tripoli. The personnel who had been in the composite troop manning the Italian 90 mm guns returned to the Regiment as these guns were handed over to a New Zealand unit, 51st HAA Regiment was relieved by 61st HAA Regiment, and 38th LAA Regiment replaced 34th LAA Regiment.

On April 13th at 7.30 p m a force of more than twelve aircraft approached from the east. This proved to be 'a very half-hearted affair' and a heavy barrage of 2,653 heavy rounds plus almost 9,000 light rounds persuaded the attackers to break off. A few bombs or mines were dropped but no damage was caused. No hits were claimed although the previous day 24 Battery had claimed damage to a Ju 86 at 28,000 ft.

Another recce plane was engaged on April 15th but it retreated quickly when fire was opened. A low flying intruder was similarly chased off in the early hours of April 26th. Further recce planes were engaged during May and June; a barrage was fired to scare off an intruder on May 31st but a recce plane at 32,000 ft on June 29th was not engaged.

On May 12th General von Arnim surrendered and the war in North Africa was over. Almost 250,000 Axis prisoners were taken at the end of a campaign which had begun in October at El Alamein and witnessed the greatest advance by a British Army since the days of Wellington.

The next step in the Mediterranean theatre would be the invasion of Sicily — Operation 'Husky' — followed by the invasion of Italy — Operation 'Avalanche'. 9th Regiment would have no part in the first other than protecting Tripoli but would play a significant role in the second.

As far back as February they had begun to train as dual purpose gunners — to use their 3·7s as field artillery in support of the ground troops. Practice shooting had been carried out at Azizia, south of Tripoli, and mobile exercises had also been executed. The Regiment's efficiency had impressed the instructors and observers at Azizia and later, in May, when practising anti-tank shooting Sgt. Jim Peoples' 3·7 had hit the target (the turret of a dug-in tank) with its first round. Although they would never fire at close range at tanks they would experience field gunnery in Italy where their guns would prove extremely effective.

At Tripoli on April 11th 25 Battery learned that two of their men with MAC Troop, Bdr Neil Murray and Gnr Thomas McCarter had been killed in action at the Battle of Wadi Akarit on April 6th. 26 Battery also learned that one of their number, Bdr Eric Rutherford had died of wounds on March 31st. Two other 26 Battery men were seriously wounded, Gnrs Fred Barrie and John Kerr, the former having an arm amputated.

The Regiment was now told that they had been earmarked for "dilution" by Basuto personnel, about 80 of whom would be attached to each Battery. The Basutos were members of the Army Pioneer Corps and they would remain on the South

African army establishment while posted to 9th Regiment. Personnel from the Regiment were sent back to Egypt to commence training of the Basutos. When they arrived and were attached to the Batteries an equal number of the existing personnel would become surplus. These men were then posted to form the nucleus of a new Battery — 653 HAA Bty/193 HAA Regt. It was an unpopular move. Since the beginning of the year it would mean that the Regiment had lost over 100 men from each Battery and was in danger of losing its identity. However, there was little that could be done and the Basutos arrived on May 30th when they commenced training on the gun-sites.

It is to the credit of the "Derry Boys" that the Basuto soldiers were accepted and made welcome in the Regiment. This "dilution" scheme, as it was officially termed, created many problems not least of which was a language barrier, for many of the Basutos spoke no English at all. The fact that they remained as Pioneer Corps personnel and on the South African establishment meant administrative problems and in their early days with the Regiment they did not even share the same food retaining their African diet of "mealie". In terms of religious affiliation there were three basic types of Basuto — Catholic, Anglican and, as one NCO termed it, "tree-hanger". Even there they retained their own Padres, serving not just the Basutos of 9th Regiment but those in other units in the Tripoli area.

To the amazement of the men of 9th Regiment the Basutos disliked the heat of North Africa. Since their own homeland was high-lying they did not normally have to tolerate the degree of heat to be found in Libya. How they would have reacted to the hothouse Sudan was a matter of speculation in 25 Battery which, like the other two batteries, received about 80 Basuto soldiers. An additional 28 Basutos were posted to Regimental HQ giving the Regiment a total of 265 African soldiers.

Before joining the Regiment the Basutos had received training at the Royal Artillery Base Depot as gun loaders. That training had to be expanded upon at Tripoli to involve them in a wider range of duties. And so during June and July the Basutos were gradually assimilated. Alongside them the personnel being replaced were training for their new role as the cadre battery of 193rd HAA Regiment. By the end of June satisfactory results were being reported from African gun teams in co-operation with 'cadre' Command-Post teams in practice shoots. Then on July 26th the 'cadre' personnel left for Base Depot. Major Robertson-Smith of 26 Battery went with them as the Battery Commander of 653 Battery and his place was taken by Captain J H McVeigh of 25 Battery. McVeigh was an original Supplementary Reserve Officer of 25 Battery, and thus the command of two of the Batteries now rested on the shoulders of Derry men — Major Sir Basil McFarland and McVeigh.

Although the Germans had been very quiet in the skies over Tripoli during this period — the reconnaissance overflights on May 31st, June 16th and June 29th being the *Luftwaffe's* only appearances — there had been plenty of activity on the ground. The batteries had undertaken a programme of mobile training, positions in Tripoli had been changed about and of course the Basutos had arrived. In addition the post of AA Defence Commander, Tripoli, had been abolished, the harbour defences now coming under the command of the AADC, Tripolitania, with the different

Regimental Commanders in Tripoli rotating the post of Sector Commander on a weekly basis. Lt. Colonel Aitken had therefore been the one and only AADC, Tripoli, a job which he had performed to a superlative degree of professionalism.

On June 1st 24 Battery acquired two 20 mm Breda guns — one for each site — which had been captured from the Italians. The following day a parade in Tripoli in honour of the King's birthday was attended by a party from the Regiment. Later in the month King George VI made a personal visit to the Middle East. He spent some days in Tripoli and all of 9th Regiment were included in parades on June 19th and 21st. Some who got close to the King were shocked to see how frail and ill he looked; his visit was therefore all the more appreciated by his soldiers.

Shortly after the King's visit the Regiment was warned to prepare to revert to a mobile rôle — only three months after they had converted from just such a rôle to being semi-mobile. They did not of course have the personnel lost then re-posted to the Regiment but at least most of the men who had gone to MAC Troop had returned in late May. In mid-July a party under Major Grant, the second-in-command, went to Cairo to draw operational equipment for the Regiment's new rôle, returning on July 30th. Operational Order 23 issued on July 27th informed the Regiment that it would be relieved at Tripoli by 69th HAA Regiment on 1st August and would pass from the command of AADC, Tripolitania to X Corps and 12th Anti-Aircraft Brigade. Accordingly on 1st August the Regiment in its entirety moved to a concentration area four miles east of Tripoli and became once again part of 8th Army.

The next six weeks were spent in intensive preparation of both personnel and equipment for the invasion of Italy. Monday 2nd August saw them begin training in loading their guns and vehicles on to Tank Landing Craft — using dummy LCTs marked out on the ground! A wireless detachment from 12th AA Brigade was attached and wireless trucks issued to the batteries — as were anti-tank rifles. Combined Operations Training, radio-telephone procedure training, maintenance and waterproofing of guns, vehicles and equipment all helped the weeks before the attack on Italy to pass quickly. By the end of August all of the guns had been waterproofed and further gun drill training had to be carried out on the guns of another regiment. All of the Regiment's personnel had been trained in both dryshod and wetshod landing and in loading and unloading their guns and vehicles on LCTs. The training on dummy LCTs proved to have been useful when it came to training on the real thing at Tripoli docks.

By now all three Batteries were commanded by Supplementary Reserve Officers who had joined the Regiment on formation, with Major J D A Henderson, a Fermanagh man, taking over command of 24 Battery on 24 August from Major Stockman (who had been admitted to hospital the day before). In 25 Battery Lieutenant E H 'Sonny' Babington was promoted to Captain — a promotion he had missed out on some months earlier due to an accident which left him with a broken pelvis.

The Regiment was to go into Italy after the main invasion force and was part of X Corps Reserve, which also included 7th Armoured Division (The "Desert Rats") with which 25 Battery had served at El Daba in 1940. X Corps however was no longer part of 8th Army but had been moved into a new army under the command of an American general — Mark Wayne Clark. This 5th Army consisted of X (British) Corps and VI (American) Corps and had been set up in Morocco earlier in the year; it held the distinction of being the first Anglo-American army since the Indian Wars prior to the American War of Independence. Among the American troops in VI Corps were men who had been based in Northern Ireland — many of them in and around Londonderry — and ironically one of the divisions of VI Corps was a 36th Division. (Derrymen had served in the 36th (Ulster) Division in France during the First World War.)

When the main invasion force sailed on September 3rd Lt. Col. Aitken went with them to take charge of the AA defences of the beachhead at Salerno. The bulk of the Regiment, which was to move in with X Corps Reserve, spent some more time training before being allotted to their Landing Ships on September 9th. Next day the Regiment was placed on two hours notice to move, with 24 Battery on 10 minutes notice. They were standing by to load on the 11th and readiness was set at 10 minutes for everyone the following day.

At 8.30 a m on September 12th the guns and GL sets were moved down to the docks and loading began. 26 Battery had completed loading by 2.00 p m and 24 Battery by 6.00 p m. However RHQ and 25 Battery were delayed due to accidental damage to their vessel — LST 200 — which had been sustained the night before. A naval repair party had to cut the doors open and by the time the task had been completed it was dusk and 25 Battery had to load in the dark. However, this operation went without a hitch and was complete by midnight. Next morning, at 6.30 a m the Regiment, less a party of 5 officers and 63 men under the Quartermaster, sailed in a convoy of twelve LSTs escorted by three Motor Torpedo Boats. The ships sailed in two lines, each of six vessels, and personnel on board were issued with lifejackets and assigned to liferafts. Bren guns were positioned on the decks for anti-aircraft defence.

At midnight the convoy passed through the Straits of Messina between Sicily and the toe of Italy and the voyage continued towards their destination — Salerno Bay — next day. There was a short 'action stations' at 10.00 a m but the 'all clear' was soon sounded and the soldiers settled down to await their arrival in Italy.

The convoy entered Salerno Bay at 6.00 p m on September 15th to be greeted by a *Luftwaffe* air attack. However the aircraft were driven off by naval ack-ack fire as the LSTs moved towards the shore. Two Royal Navy battleships, *Warspite* and *Valiant*, were shelling enemy shore positions while German 88s were "dumping" a frightening barrage on the beachhead. It was, according to Robert Hamilton, a frightening scene and there were many who doubted if they would ever see home again.

As the LSTs neared the shore the troops were told to "hold on tight" in readiness for a heavy landing. Instead, recalls John Ormsby, there was a barely perceptible bump as LST 200 grounded; then the doors opened, the ramp went down and 25 Bty disembarked at 9.00 p m without getting wet.

24 Battery's ship had made a number of attempts to beach, being warned off

each time because of mines, before going ashore at 10.30 p m. Unloading, again dryshod, was complete within an hour.

26 Battery beached at 9.00 p m and although there was no beach organisation in sight began to unload. The Battery Commander's recce car was driven off first and manhandled along the soft beach followed by two X Corps RASC lorries (loaded at the last moment at Tripoli) which promptly sank into the sand. Just then an officer from the MLO (Military Landing Organisation) arrived in an amphibious DUKW and ordered the LST off again. The beach they had landed on was mined! Major McVeigh's recce car and the two RASC vehicles were abandoned — the latter complete with crews. McVeigh's car was later retrieved. 26 Battery finally went ashore on Sugar Amber beach at 2.30 a m on September 16th.

They had arrived at a critical phase in the operations at Salerno. A strong German counter-attack threatened X Corps which had only one battalion of infantry (Grenadier Guards) in reserve. 9th Regiment were about to become infantrymen in order to prevent Operation Avalanche from becoming another Dieppe.

CHAPTER SEVEN

THE AFRICAN ORANGEMEN

Many of the warrior races against whom the British army has fought across the face of the earth have become close friends of Britain. In many ways this friendship was a tribute to the respect which those peoples had for the British soldier, and indeed that friendship has often extended to fighting alongside British soldiers. Perhaps the best known of these races are the Gurkhas of Nepal and the Sikhs of India. Not so well-known, either as adversaries or as allies, are the Basutos of southern Africa. In the second half of the nineteenth century the Basutos had fought the British as valiantly as did their Zulu counterparts. Briton and Basuto recognised each other as worthy opponents, and a mutual respect and bond developed.

When South Africa entered the second World War on the Allied side her white soldiers were soon augmented by coloured troops among whom were numbered the Basutos. At first, to their chagrin, the Basutos were not given combat tasks but this began to change and the smiling tribesmen from the Orange Free State soon began to take their place in fighting units.

Basuto troops came to 9th (Londonderry) HAA Regiment in May 1943. That their arrival was unpopular was in no way due to any prejudice about the colour of their skins but rather to the fact that their arrival meant that many men who had been with the Regiment since 1939 would have to go. Precisely the same attitude would have prevailed had the newcomers been English, Scots or Welsh. The feeling of the "Derry Boys" was that the army was trying to destroy the identity of the Regiment in spite of having given assurances in '39 that that identity would be preserved.

The attitude of the Basutos themselves soon broke down any reserve which existed. Before long these smiling African highlanders were proving themselves to

A Basuto soldier of 9th Regiment. This photograph of Pte Lekula was taken in Italy in October 1944 after he and his comrades had transferred to 55th HAA Regiment.
(Photo: Imperial War Museum)

be capable gunners and not simply carriers of ammunition. Jim Canning remembers them as extremely good and willing soldiers who were a joy to work with.

The Basuto diet did however cause some concern according to Sonny Babington. At first they existed on a regime of "mealie" which seemed to induce corpulence rather than solid muscle. In Italy their rations were changed to standard British rations with the by then usual addition of some American items. The change-over was not something to which the Basutos took kindly at first. A number of ex-gunners reckon that this was because the "mealie" could be, and indeed was, used to produce an alcoholic brew and some Basutos are said to have believed that the change was intended to deprive them of their "mealie" beer.

Numbers of Basutos produced their own messtins by the simple expedient of procuring a large empty syrup tin and having a "tiffy" fix a handle to it. They would then line-up at mealtime and collect meat, vegetables, potatoes, "duff" and tea in the one tin at the same time. The resultant Basuto stew was an unbelievable concoction to the "Derry Boys", although it wouldn't be too strange to anyone who has served in the Royal Navy.

Many friendships grew up between Basuto and "Derry Boy". Sgt. Raymond "Chunkie" Semple and Griffith Molipa maintained correspondence long after the Regiment left Italy where it said goodbye to its Basutos. Molipa sought advice from Semple on what he should do after the war as well as passing on greetings from other ex-9th Regiment Basutos.

There are many happy memories of the Basutos — such as Joseph serving up tea at Salerno with a "smile as wide as a map of Africa" and of laughing Basutos wringing out their soaking blankets after a night of torrential rain at Salerno, of the Basuto who came to Captain Sonny Babington to become a Catholic and to change his first name to Joseph! Babington remembers too the distress shown by a Basuto who had received bad news from home — someone was stealing the hair from his Angora goats and to a Basuto that was worse than stealing a wife!

There was tragedy, too, for of the fifteen men killed at Naples on October 21st 1943 ten were Basutos. Men like big Sgt Choni Kharasi, Cpl Solomon Mosestheli or Pte Tsoebilane Lepota were mourned as keenly as their white colleagues, for they too had become "Derry Boys" in their time with the Regiment.

A Vision of Hell! Guns of 25 HAA Battery firing in a counter-battery role in the Salerno beachhead.
(Photo: Imperial War Museum)

A 3·7 of 9th HAA Regiment passing through Torre Annunziata, 1st October 1943. The Battery is not known but the gun belongs to a B Troop.
(Photo: Imperial War Museum)

PART FOUR

THE SOFT UNDERBELLY

CHAPTER ONE

A VISION OF HELL — SALERNO

The circumstances which brought the men of 9th Regiment on to the beaches at Salerno that September night in 1943 were created by a combination of factors — political and military. Stalin had been pressurising the western allies for the opening of a "Second Front" to ease the pressure on the Red Army. The Americans, pouring men and equipment into the United Kingdom, wanted to attack across the Channel in 1943. Churchill, supported by Brooke, did not want an invasion during 1943 believing that the combined British-US forces could not sustain such an operation — and the lesson of history is that his judgement was sound.

Moreover, opined Churchill, a "second front" had been in operation in North Africa and therefore he proposed using the forces in that theatre for an attack on what he described as the "soft underbelly" of Europe — Italy. This idea was accepted by Stalin and Roosevelt and plans were drawn up for its execution under the overall command of General Dwight D Eisenhower.

As a preliminary to the invasion of Italy, allied forces — in the form of 8th Army and the US 7th Army — landed in Sicily on July 10th 1943 in Operation "Husky". The campaign was concluded by mid-August although, unlike North Africa, where some 250,000 Axis prisoners were taken, the bulk of the German and Italian forces escaped across the Straits of Messina.

Secret negotiations were taking place with the Italian government — Mussolini had been overthrown in July — for an armistice and this was concluded on September 3rd but not announced publicly. On that same day General Montgomery's 8th Army, under artillery and air cover, crossed the Straits of Messina and landed on the "toe" of Italy. This manoeuvre, Operation "Baytown", had two objectives — the first to secure the straits for allied shipping, and the second to have 8th Army move up from the "toe" to join up with 5th Army as soon after the latter landed at Salerno as possible. As an added "distractor" for the Germans the "Avalanche" landings at Salerno would be accompanied by a simultaneous attack at Taranto — Operation "Slapstick".

The area selected for the Salerno invasion was tailor-made for such an operation. South of Salerno itself are about twenty miles of wide beaches with no natural obstacles. Of course, that this was ideal as a landing site was as obvious to the Germans as it was to the Allies. Although Clark would have preferred an invasion north of Naples the air forces could not have given adequate cover and the Italian request for landings south of Rome was impossible for the same reason. Salerno

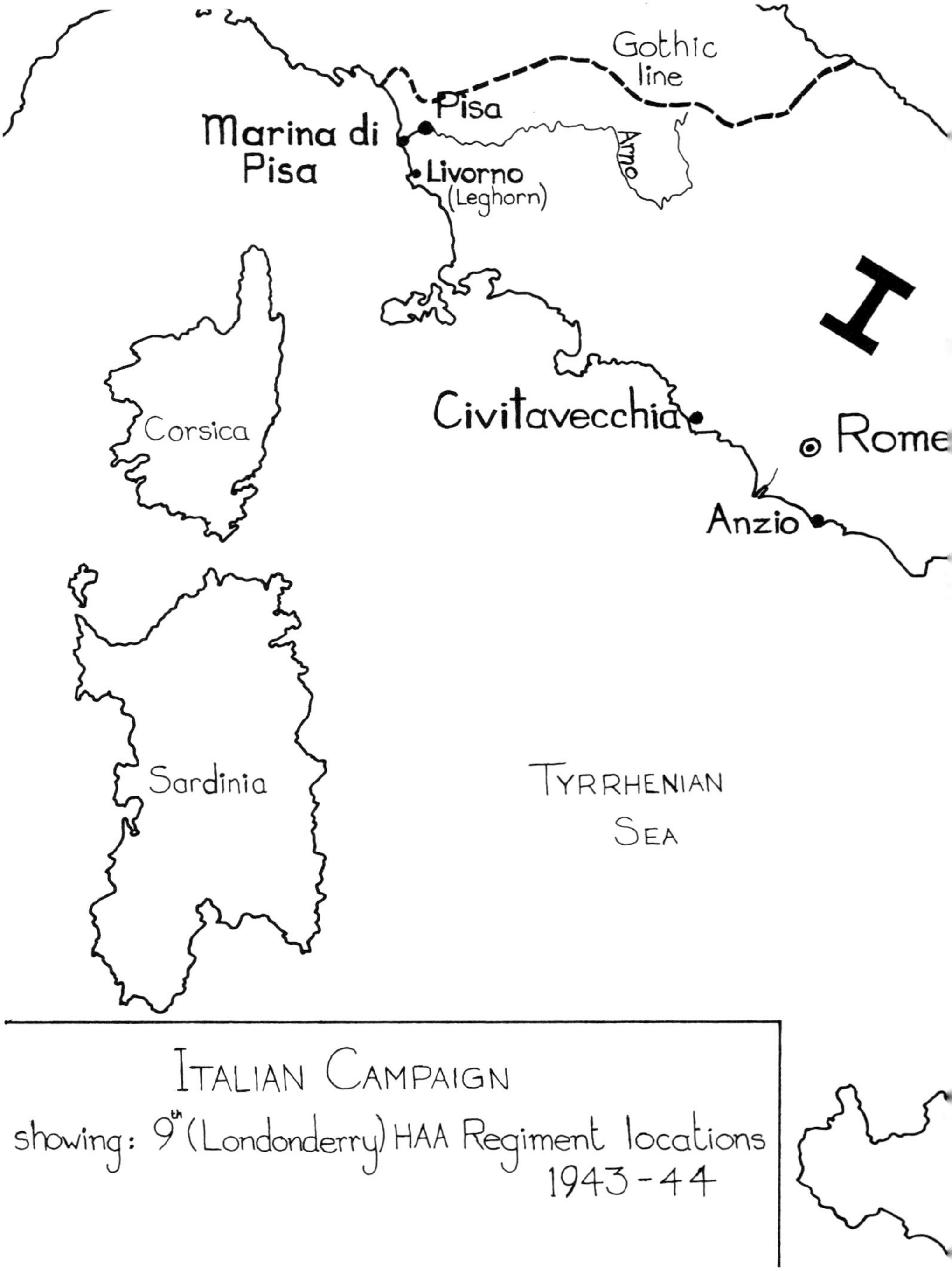
Gothic line
Pisa
Marina di Pisa
Livorno
(Leghorn)
Arno
I
Civitavecchia
Rome
Anzio
Corsica
Sardinia
TYRRHENIAN SEA
ITALIAN CAMPAIGN
showing: 9th (Londonderry) HAA Regiment locations
1943-44

YUGOSLAVIA
ADRIATIC
SEA
ALY
Naples
Torre d. Greco
Vesuvius
Castellammare
Salerno
Torre Annunziata
Ravello
SICILY

however was within range of fighters based in Sicily and thus one could almost say that Salerno selected itself.

While the first wave of invading troops was at sea between North Africa and Italy the surrender of that country was announced. German forces moved immediately to "neutralise" the Italian army and to take over the defences of the country. Against the advice of Rommel, who was now in command in northern Italy and who felt that Italy should only be defended north of Rome, Hitler ordered that the invaders should be forced to fight every step of the way up the peninsula. Thus as 5th Army moved in to land General von Vietinghoff's 10th Army was deployed to resist them.

The first assaults were launched on September 9th by the British 46th and 56th Divisions of X Corps in the northern sector of the Gulf of Salerno and by the American 36th (Texas) Division in the southern sector. Resistance was light at first as Vietinghoff held back his forces until he was certain that Salerno was the point of maximum effort and not a diversion. Then he concentrated two *Panzer* and two *Panzer Grenadier* Divisions against the beachhead in accordance with his Supreme Commander Field Marshal 'Smiling Albert' Kesselring's directive that "the invading army . . . must be completely annihilated and, in addition, thrown into the sea".

Montgomery's 8th Army found their progress hampered by extremely effective demolitions by the retreating Germans and by small battle groups which engaged the British troops as they advanced through the rugged Calabrian countryside. There was no chance of 'Monty' being able to link up with 5th Army as planned.

Vietinghoff's troops began to press hard on 5th Army and to exploit the inexperience of the US 36th Division. Clark's ability as a Field Commander was also severely put to the test during the next few days, and indeed at one point he considered evacuating the American troops from the beachhead and putting them ashore again in the British sector. This idea was rightfully dismissed by his immediate superior, General Sir Harold Alexander, who commanded 15th Army Group to which both 5th and 8th Armies belonged, as it would undoubtedly have created panic among the inexperienced American troops. Alexander inspected the front line situation for himself and promised Clark the use of the US 82nd Airborne Division whenever and wherever he (Clark) wanted it. 1,600 men of the Division dropped on to the beaches on September 14th and provided the vital steadying power for the 36th Division which had by now been joined by the 45th Division.

September 13th had been the day on which Vietinghoff launched his main thrust at the 5th Army front. Four regimental-sized battlegroups advanced over a nine-mile front against the British right flank and the American left flank. Only by putting every available man, including cooks, mechanics and stretcher-bearers, into action as infantry did the Americans prevent the Germans from breaking through. Next day another strong German attack concentrated against the Americans was held, and when at night-time the Hermann Goering *Panzer* Division attacked the British 46th Division they too were driven back. The situation was still critical but the arrival of the American paratroopers helped to stabilise it.

Fierce fighting continued throughout the next day, and by the night of Wednesday

September 15th the British X Corps had no reserve of infantry left. It was into this fraught situation that 9th (Londonderry) Regiment landed between 9.00 p m on the 15th and 2.30 a m next morning.

At midnight on September 15th the Regiment was ordered to assume an infantry rôle. By 2.00 a m Ninth had become a Corps Reserve of Infantry. 468 Officers and men were formed into Rifle Companies, and placed under the command of Lt Col Browning of 52nd LAA Regiment and moved to take up positions near Montecorvino aerodrome which was held by X Corps troops.

From these positions, at Campo di Fortuna Caputo, patrols were sent out at 6.00 a m that morning. Moving along the railway line about 1,000 yards south-east of Campo di Fortuna Caputo one patrol encountered a small group of infiltrating German infantry. The gunners fired on the Germans, who withdrew leaving one of their number dead as a result of the engagement. Another patrol consisting of men from 26 Battery caught a German sniper unawares and shot him. This may have been the same sniper who, operating from a railway signal box, opened fire on a 25 Battery gunner who had, with due deference to modesty, moved into the cover of some bushes to relieve himself. He wasn't hurt but he did rapidly retreat to rejoin his comrades beside the railway embankment!

Rifle and machine-gun fire continued throughout the morning together with shelling and mortaring. One of the most frightening aspects of the shelling was the fire from 88 mm anti-aircraft guns being used as field-guns. Their shells were fuzed to 'air-burst' above the ground showering red-hot shrapnel over the soldiers beneath. Even under such bombardment the "Derry Boys" kept their sense of humour. As his comrades dived for the cover of slit-trenches one gunner was heard to remark that there was no point in taking cover as "they follow you in". One man who dived quickly into a slit-trench found that he had in fact landed in a latrine-trench which had not been filled in! "Any port will do in a storm" was his reaction; nonetheless his companions stayed well away from him for some time!

Some of the shells were aimed at the concentration areas where the Basutos, because of the language problem, plus some other men from each battery had been left to guard the guns and equipment. No casualties were caused however.

A tank attack during the morning was repelled by infantry anti-tank gunners, and at 10.00 a m the Rifle Companies moved up to more forward positions still under enemy shelling. 25 Battery moved an hour later to become lorried infantry and were attached to 1st/6th Battalion the Queen's Royal Regiment (Queens). By 2.00 p m all three Battery Rifle Companies were attached to 1st/6th Queens. This Regiment, part of 131st Brigade, had landed as a reserve unit but only their Battalion and Company Headquarters had arrived. Accordingly 24 Battery became A Company, 25 Battery C Company and 26 Battery D Company, 1st/6th Queens. Officers from the Queens were attached to each company as infantry advisers. All three Battery/Companies were "lorried" and moved to the Incarto area.

There was a "stand-to" at 5.15 a m on the Friday morning (September 17th) and the gunners took up defensive positions. No action resulted and at 7.00 a m 1st/6th Queens received their own personnel and moved south. 5th Royal Tank Regiment became the next "home" for the men of 9th Regiment. That afternoon the personnel of 25 Battery were withdrawn from the infantry rôle and ordered to deploy with their 3·7s for the anti-aircraft defence of the beaches. Throughout that day 24 and 26 Batteries remained in a reserve infantry role with 5 RTR and at 6.00 a m on the morning of Saturday, September 18th they were re-assigned under command and in support of the Royal Scots Greys.

With the pressure on the beachhead easing, due to intense naval bombardment and to attacks by allied aircraft plus the proximity of Montgomery's 8th Army, the rifle companies were disbanded on the Sunday morning at 8.30 a m and the personnel of 24 and 26 returned to their regular role.

In the meantime 25 Battery had been placed under the command of 100th HAA Regiment and A and B Troops had occupied gun-sites which had been dug by bulldozers — the first occasion on which the men had not had to dig the pits themselves! The sites had been visited by the AA Defence Commander, Salerno, who was none other than their own CO, Lt Col Aitken, and A Troop had fired 11 rounds at hostile aircraft which had dived on the beachhead from out of the sun. On Sunday they had fired on another aircraft approaching from out of the sun but it had been recognised as a Bristol Beaufighter (used by both the RAF and USAAF) and the engagement ended. There were strafing attacks on Monday and Tuesday, September 20th and 21st, and both Troops were in action. On Monday morning at 7.00 a m A Troop had fired 9 rounds at five Bf 109s and at 6.15 p m B Troop put up 23 rounds at two Bf 109s. Both Troops fired on attacking 109s at 10.55 a m on Tuesday and were again in action simultaneously at 7.30 a m on Wednesday.

After resuming the anti-aircraft rôle 24 Battery had been ordered to fire only by night. They had in fact been forced to delay occupying their positions until after dark on the Sunday due to shelling, but both Troops were ready for action by 10.50 p m. Intermittent shelling continued throughout the afternoon and evening of Monday, and on Tuesday morning the Battery was told to prepare to move to relieve 213 HAA Battery. The shelling which had delayed 24 Battery from occupying its sites also forced Regimental HQ to move to a new location at 3.00 p m on the Sunday. This was in fact heavy enemy counter-battery fire on nearby medium guns.

Since re-assuming their anti-aircraft duties 26 Battery had been defending RAF landing sites in the beachhead area. They were also in action against enemy fighters. At 6.07 p m on Monday September 20th five or more Focke-Wulf FW190s dived through clouds to attack the landing-strip. Several bombs fell near the guns and A Troop fired 7 rounds in return. Another 9 rounds came from B Troop, and the former also fired some 200 rounds of machine-gun fire. At 10.45 next morning a flight of FW190s came in at low level. A Troop fired 14 rounds and B Troop 3 rounds plus 100 rounds of machine-gun fire. One of the FW190s engaged by A Troop was seen to crash into the hills around the bay and explode. FW190s launched a low-level "tip and run" raid on Montecorvino airfield at 7.45 a m on September 22nd. Bombs were dropped as the fighter-bombers made off at high speed. B Troop did not see the raiders but A Troop fired 13 rounds.

At noon that day 25 Battery had moved into an 8-gun position which had been occupied previously as a 4-gun position by 213 Battery. They were now deployed as field gunners in a counter-battery role. In this rôle they were joined later that day by 24 Battery, while 26 Battery remained as ack-ack gunners at the RAF landing-strips.

As this was happening the A Echelon which had remained in Tripoli arrived at Salerno and disembarked to move into X Corps' concentration area. From there they moved up to rejoin their respective Batteries. (B Echelon remained in Tripoli and would not sail until September 26th.)

24 Battery fired its first rounds in counter-battery work in the early hours of September 23rd. This was part of what was described as a Corps "Neutralisation programme" (i.e. it was intended to put enemy artillery out of action). Firing at the rate of seven rounds per gun per minute the Battery fired a total of 566 rounds most of which (398) came from B Troop. A shell bursting close to one of 24's positions damaged a gun limber and a bantam car, and splinters from these wounded six gunners, four of them Basutos. Two Basutos had to be taken to a Casualty Clearing Station but the other men had received less serious wounds and remained on duty.

25 Battery engaged in similar fire, their orders being to put down harassing fire on six specific targets at ranges from 11,300 to 14,400 yards. These targets included both guns and armoured vehicles and the fire proved extremely effective.

Both 24 and 25 Batteries remained in the field artillery role for a week, by which time 5th Army had effected its break-out from the beachhead area, and the crisis which had threatened to turn Salerno from the greatest amphibious landing to date into another Gallipoli or Dieppe had passed.

A move by A Troop/25 Battery at 11.00 a m on the 25th had to be abandoned as the new position immediately came under enemy fire. Fortunately, there were no casualties and A Troop returned to its previous site. B Troop/25 Battery were able to take up a new position without interference. 24 Battery also came under enemy fire that day with a number of massive 170 mm shells falling across B Troop's position. At 5.30 p m that evening 24 and 25 were again involved in counter-battery fire with 24 firing 387 rounds. During the evening 24 Battery had another casualty from a shell splinter but the wounded man remained on duty.

During the night of September 26th/27th both batteries again laid down counter-battery fire. The War Diary of 24 Battery records 252 rounds fired by A Troop and 246 by B Troop. Between 9.00 p m on 26th and 3.30 a m on 27th the guns of 25 Battery fired 238 rounds at six specific targets at ranges between 8,100 yards and 13,900 yards. The guns ceased fire before dawn — the Germans were beginning their withdrawal and the battle of Salerno was over. In a Special Order of the Day General McCreery commanding X Corps said

> "After ten days of heavy fighting Tenth Corps has forced the enemy on to the defensive and on our right he is carrying out a big withdrawal. His object was to drive us back on to our beaches . . . He has failed — thanks to the courage, endurance and splendid fighting spirit of you all. You have inflicted very heavy losses on the enemy and he has lost much equipment . . ."

While 24 and 25 had been firing as field gunners 26 Battery had remained in their defensive role near Montecorvino airfield. After the attack by FW190s on September

22nd there followed two days without any shooting in their sector, but at 9.27 a m on September 25th they were back in action again as four aircraft — one of which was a Bf109 — attempted to bomb and strafe the road near the airfield. The aircraft made off rapidly when the guns opened up. One machine — the Bf109 — was seen diving from clouds apparently on fire and may have been damaged by gunfire or fighters. At 12.25 an aircraft was spotted diving from about 3,000 feet to 1,000 feet. It was out of range to the guns however and when last observed was being pursued by three allied fighters.

On September 26th there was a Battery Commanders' Conference at Regimental HQ at which the plans for the anti-aircraft defence of Naples were discussed and Battery sites were allocated on the map. 26 Battery's Commander, Jack McVeigh, who had experienced difficulty getting to the conference due to the congested nature of the roads (he made it on the pillions of various motorcycles and by walking) noted that a recce of the sites "will be a bit difficult". That was a slight understatement — Naples was still occupied by the Germans!

Finally on September 29th came the order to move out of the beachhead area. That evening the Regiment moved into Salerno town where the men bedded down for the night on the streets. One gunner of 26 Battery became a casualty when he fell into a bomb crater in the dark and had to be admitted to hospital. After a leisurely breakfast next morning the Regiment set off for a concentration area near Angri only 15 miles away. It took over ten hours to get there, such was the volume of traffic on the roads. Late that night they were informed that they were to leave Angri next day to take up positions at the little harbour of Torre Annunciata on the Bay of Naples.

The Regimental convoy arrived in Torre Annunciata on October 1st, as 5th Army troops were entering Naples, and deployed under the command of the AADC, Torre Annunciata. Several hours later came the order to move again — into Naples for the defence of Naples harbour. Leaving Torre Annunciata at 5.30 a m next morning they moved to Naples and by noon were operational in six 4-gun sites with Regimental HQ in the Observatory. Lt Col Aitken was AADC, Naples and three light batteries also came under his command.

German snipers were still firing from buildings in the City as they moved their guns and instruments into position. Standing orders for the defence of Naples were issued almost immediately, together with data for the various barrages which would be part of that defence. Over the next few days a further HAA Battery — 215 — and another light battery — 195 — together with a Smoke Company were added to the defences. All was ready for whatever the *Luftwaffe* in Italy could throw at Naples.

Since landing at Salerno, just over two weeks before, the Regiment had seen its fair share of action. Under fire for most of that time it had acquitted itself admirably as infantry and field gunners as well as in its normal role. Casualties had been slight with no deaths or very serious injuries, even though units close to them had suffered badly. Throughout the trauma of those days in the beachhead the gunners had displayed a calm unruffled attention to duty that commanded the admiration of all who came in contact with them.

This is perhaps best illustrated by an experience Bertie Cuthbert of 25 Battery had many years later when attending a meeting in Belfast. Held in a government establishment this meeting was dealing with business investment in Northern Ireland. Bertie arrived early and was ushered into a room with another early arrival and offered tea, which was accepted. When the refreshments arrived he was amazed to find a mug without a handle on the tray and he took it with the comment 'like being back in the army'. The other man commented that he had been in the army too and said that he had very clear recollections of the best cup of tea he had ever had. Indeed, he said the story of that 'cuppa' would interest Cuthbert as he came (did he not?) from Derry.

> This man had been at a place called Salerno during the war — "you might have heard of it". Anyway it transpired that he and a companion were taking cover close to a railway embankment during one night of the operations there. During that night the air had been filled with the sounds of shell fire, mortars, machine-gun and rifle fire. During quiet spells the pair could clearly hear a group of men singing. One song they recognised as being "The Sash" and reckoned that the singers had to be Inniskilling Fusiliers. Various other Irish airs could be heard throughout the night.
>
> As dawn approached the dew began to fall and it became bitterly cold, the cold seeping through all their clothing and into their bones. It was just after first light when the pair noticed that the group of singers were brewing tea, sitting around their "brew" without steel helmets while shells passed overhead! Both men were longing for a hot cup of tea and eventually called to the other soldiers to ask if they could have a mug each. On receiving an affirmative reply the storyteller began to crawl across only to be told in a Derry accent "get up out of that, man! A wee mortar'll do you no harm". He walked the last few yards and although he admitted he was terrified, walked back to his mate with the tea. It was, he declared, the best tea he had ever tasted.
>
> But, he added, the soldiers hadn't been "Skins". Indeed they had been some regiment of gunners from Derry and they had "a big shot" with them. "A big shot?" queried Cuthbert, "there aren't too many of them in Derry. What was his name: McCorkell?"
>
> "No" "Gillanders?"
>
> "No" "Watt?"
>
> "No" "McFarland?"
>
> "McFarland, that's the name — he was their Battery Commander"
>
> "That was my Battery", said Bertie Cuthbert, "but, tell me, what was our reputation?"
>
> "You were called the mad-hatters! Nothing seemed to frighten you".
>
> "By the way, what were you in yourself?"
>
> "Royal Marine Commandos".

Coming from a Marine that was some tribute.

First Sergeant Ben McKelby, (centre) US 82nd Airborne Division pictured at Naples with Sgt John Taylor (left) and Gnr Jack Hunter of 24 HAA Battery, October 1943.
(Photo: John Taylor)

Sgt James Soutar, Sgt Jim Canning and BSM Walter Wray of 25 HAA Battery in Italy, 1944.
(Photo: Sir Basil McFarland's Collection)

CHAPTER TWO

STAR-SPANGLED NEIGHBOURS

Fifth Army is remembered as an American army and it is now little known that it began its existence in early 1943 as a joint Anglo-American force. Composed of two corps, one American and one British it was destined for service only in Italy. Initially Fifth Army was made up of the US 6th Corps and the British 10th Corps but as the Allies fought their way up through Italy the US 2nd Corps was added, as was the French Expeditionary Corps, and by the time Rome was liberated 10th Corps had transferred to its "spiritual" home of Eighth Army.

However Fifth Army still contained British troops — now known as British Increment, Fifth Army — including 9th (Londonderry) HAA Regiment. The "Derry Boys" in fact remained with Fifth Army until their final relief in Italy and it was only natural that they got to know their American allies extremely well.

At first the British soldier's impression of his trans-Atlantic comrade-in-arms was not good, and this was hardly helped by the decisive defeat suffered by a numerically superior American force at the hands of Rommel's troops at Kasserine Pass in February 1943. As time passed the GIs became battle-hardened however and fought well and bravely in Sicily with Patton's Seventh Army.

In spite of the misgivings of their commander, General Mark Clark, the men of the US 6th Corps acquitted themselves well at Salerno when one remembers that they were, in the main, untried troops. The presence of the experienced British troops of 10th Corps, plus the Royal Navy's bombardment and the intervention — on General Alexander's orders — of the US 82nd Airborne, prevented 6th Corps from being overrun and Clark from the possible fate of being consigned to the history books as a military leader in the mould of General George Custer!

It was at Salerno that the Ninth's gunners had their first encounters with Americans in a combat zone. 24 Bty's men could not believe that among the *impedimenta* that the US force had brought ashore was a mobile laundry, but, in the words of Robert Hamilton, "we soon made use of it".

An American Smoke Company was in operation on the beachhead near 25 Bty and men from both units soon struck up friendships; in fact the Smoke Company was later to come under the command of the Regiment in an air defence role in the Naples area.

But the gunners were amazed to see what they often regarded as extravagance on the part of the Americans. At one location 25 Bty had a US anti-aircraft artillery (AAA) battalion (the US equivalent of a Battery; the US army Battery equated to a Royal Artillery Troop) as their neighbours. As the GIs moved in one 25 Bty man was heard to exclaim, "they must have a motor for every man!" In its mobile role "Two-Five", in common with its sister batteries, had to make do with fourteen lorries, a couple of light cars or jeeps and a handful of motor-cycles. Their counterparts employed more than forty vehicles for the same job! The "Yanks" brought their comforts with them — including mobile shower units which they willingly allowed the Derrymen to use also.

In the city of Naples the AA defences were a British responsibility but the harbour itself was under the control of the Americans. Thus when 25 Bty took over a large building in the city for its BHQ they had no sooner installed themselves than the Americans demanded use of the entire building. However an amicable compromise was reached with 25 remaining as lodgers and friends of the principal tenants. 24 Bty were also cheek-to-jowl with US troops. A Troop/24 Bty shared a sports stadium with RASC personnel and with men of the 82nd Airborne Division. John Taylor became very friendly with First Sergeant Ben McKelby of the 82nd and this was but one of many friendships struck up between gunners and airborne soldiers — many of whom came from the state of Texas.

Under the stands of the *Sportivo Militare* the accommodation had been used by the Germans as an ammunition store, and this was the area which was taken over by the men of 82nd. Some of the German munitions were still in the stands and one day — due it was assumed to some soldiers tampering with the ammunition — there was a large explosion which blew up part of the stands. In the wreckage many men were trapped and 24 Bty members joined in the grim task of trying to locate and rescue survivors, although it turned out to be a case of uncovering dead bodies more than finding living paratroopers.

General Mark Clark himself became well-known to many of the gunners. Indeed the tall American developed a habit of "dropping in" on the Ninth and was quite prepared to stand and chat quite informally with the gunners. This easygoing approach, which was in marked contrast to that of most senior British officers, appealed to the "Derry Boys" and encouraged one to ask Clark why he spent so much time with the Regiment. The reply was astonishing, for Clark told them that he felt that his back was safer when he was with British troops than when he was with his own countrymen! Clark was obviously well aware of the animosity towards him after the slaughter of the Texans of 36th Division on the Rapido river (see author's note at end of Part Four, Chapter 6).

Their eagerness to get things done sometimes resulted in unnecessary excesses. One example was the occasion when a battalion of American guns was moving into new positions in a vineyard and US combat engineers prepared the way by cutting the entire vineyard to the ground! Gunners from the Ninth who witnessed this have never forgotten the heartbroken Italians who watched the work of generations being destroyed in an hour or so.

In spite of all the differences which existed between them the "Derry Boys" and the "Yanks" of Fifth Army got on well together, with the professionalism of the gunners being especially appreciated by the American troops who felt that little bit more secure when they knew that the 3·7s of the "Irish triple A[1] guys" were supporting them.

[1] Triple A:— AAA or Anti-Aircraft Artillery.

CHAPTER THREE

"SEE NAPLES AND DIE"

Naples fell to 5th Army on October 1st 1943 and Clark decided to enter the city in triumphal style as a liberator. Such behaviour was typical of him, and more than one of his fellow-countrymen described him as fighting the war with one eye always on the media. His entry into Naples, however, backfired. With the route made safe by men of 82nd Airborne Division, Clark drove into the city in an armoured half-track truck. In his memoirs "Calculated Risk" he described his disappointment at receiving no rousing reception from the people of Naples. "There was little that was triumphant about our journey" he wrote, going on to describe the deserted streets and the feeling that he was being watched from behind closed shutters, a situation and a feeling that remained as he drove out of Naples. Even in Garibaldi Square, where he had intended to deliver his victor's speech, he saw no one. For a man who conducted his campaign with a constant eye on what history would say about him he had paid little attention to the history of Naples. Traditionally conquerors had entered the city from the north and were received, if a welcome was appropriate, in the Plaza Plebiscito just a mile from Garibaldi Square. As Clark surveyed an empty Garibaldi Square thousands of Neapolitan citizens were waiting to greet him in the Plaza Plebiscito!

After the fall of Naples Alexander ordered 5th and 8th Armies to advance northwards. The Allies still expected a relatively easy progress, but Kesselring soon changed their ideas. As the Germans withdrew to the Volturno River they demolished bridges and roads and fought small delaying actions — one of which saw three German battalions hold up almost the entire X Corps in a bottleneck until other German formations could withdraw in good order. The Volturno was reached four days after Naples fell; these actions cost some 12,000 casualties, most of them British.

Kesselring held the Volturno until October 16th, when he began an ordered fighting withdrawal back to his winter line — the 'Gustav Line'. The weather also came to the aid of the Germans, with winter beginning in October instead of the more usual November. Sunny Italy was to have one of its worst winters ever and that winter would be an ally to Albert Kesselring.

The port of Naples was a valuable asset to the Allies' supply lines. Although there were other ports available at Salerno, Taranto (in the south) and Bari (on the east coast), Naples was the most convenient port for re-supply of 5th Army. Its defence was therefore of vital importance to that army, and that defence was vested primarily in 9th (Londonderry) Regiment. Under the command of Lt-Colonel Aitken as AADC, Naples, the Regiment occupied six 4-gun sites around the city which were located for the defence of the harbour. A Battery of 57th HAA Regiment — 215 Bty — was assigned to them, occupying another two sites, while three light AA batteries — 38, 100 and 155 LAA Batteries — occupied a site each. No 1 AA Operations Room was the nerve centre of the gun defence of Naples, which could also be supplemented by the guns of ships in the harbour. By October 9th 195 LAA

Battery was added to the defences, and between 8th and 14th October three smoke companies, two of them American, joined.

As yet the defences had not been put to the test. The *Luftwaffe*, a depleted force in Italy, had been busy harassing the advance of the ground forces of 5th and 8th Armies and thus the gun sites could be prepared and equipment checked. Bad weather hindered the work of preparing the sites, some of which were close to abandoned Italian army 90 mm gun positions[1]. Equipment in those positions could prove useful, and while checking out an Italian position for items usable with the 3·7s a group of gunners from 25 Battery had a narrow escape. Because of the German snipers still operating in the area Thompson sub-machineguns had been issued and were carried as a matter of routine. While the party was inspecting an Italian gun one of their number accidentally discharged his 'Tommy-gun'. One round ripped John Ormsby's trouser leg but no-one was injured. Later it was discovered that other rounds had struck some of the 90 mm shells which were still stored in the gunpit! Only luck had ensured that all of the stored ammunition had not exploded killing the 25 Battery men. Luck however can be very cruel and fickle — and proved to be both when the first heavy raid was launched against Naples Harbour on October 21st.

In the early evening of October 21st 1943 the guns were at full readiness. As yet they had fired only once — at an unidentified aircraft thirteen days earlier, when the three batteries had fired off 187 rounds — but they knew that inevitably the *Luftwaffe* was going to hit Naples and hit it hard.

A Troop/25 Battery was located at N11 site at Villa Nuova on hills overlooking the Bay of Naples. Billy Smyth was the Sergeant on duty in the Command Post at the Sports Stadium (N9 site) when the first reports of approaching aircraft came in. He ordered a telephonist to notify the Operations Room but the call was not answered.

John Ormsby and "Ginger" Evans, two close friends, were off duty and in a hut less than twenty yards from No 2 gun. John Ormsby clearly recalls that no alarm had been given even though the sound of many aircraft could be heard almost overhead.

Miles Stott was a lance-bombardier on the nearby No 3 gun, which opened fire without waiting for the alarm. The remainder of the guns opened up also, firing a harbour barrage. Such a barrage effectively put a shield of bursting shells over the vulnerable port and was a terrifying experience for aircrew. Of the formation of some fifty bombers (a large proportion of the available strength of *Luftflotte* 2 in Italy) many of the aircraft turned away, but some dozen pilots, made of sterner stuff than their fellows, pressed on.

No 3 gun soon ran into difficulties with its firing handle and intermediate firing lever, and L/Bdr Stott left his gun pit to get replacements, which were held at No 2 gun. Above No 2 gun 'Ginger' Evans had decided that the safest place to be with a raid going on was in a gun pit and he headed for No 2 gun, knowing that the crew would always welcome an additional pair of hands to feed ammunition to the gun while John Ormsby went down to the Command Post.

[1] There were other Italian 90 mm guns still operating in defence of Naples.

At No 1 gun Sgt Jim Peoples had his gun-crew working as the smooth and professional team which they were. In all such actions each crew vied with the others to see who had fired most rounds — there was no reason why this night should be any different.

John Ormsby recalls hearing the sound of an aircraft diving and it seemed to be overhead. 'Ginger' Evans had reached No 2 gun but Miles Stott was only part-way between the two guns when he was lifted off his feet and blown back in the direction from which he had come. John Ormsby can remember the sight of the huge mass of flame, debris, smoke, earth, blood, bone and metal that surged up in front of him. And yet he heard no sound, although like Miles Stott he felt the heat of the explosion.

It is easy to assume that the bomb which blasted No 2 gun and its crew was a freak occurrence; perhaps the result of a pilot jettisoning his bombs and abandoning his mission. However, eye-witness accounts provide evidence that it was in fact a deliberate attack.

Willie Sherrard, a Lance-Bombardier in 25 Bty, was returning from a day pass with a companion when the raid began. Just before the bomb hit the gunsite the two friends saw a light flashing from the roof of a house. Both were convinced that the machine which did the bombing flew in over this house which was in direct line of N11 position. In any event the guns were already in action and their position was obvious as their muzzles belched forth sixty feet long flashes of flame.

Robert Hamilton of 24 Bty remembers one aircraft darting around the barrage towards 25 Bty's positions. The respective locations of Hamilton and the two 25 Bty men make it more than likely that they all saw the same aircraft. Add to this John Ormsby's evidence of the sound of an aircraft *diving* on the position — as an experienced ack-ack man there is no doubt of his testimony — and the conclusion must be that this was a calculated attack.

The effect of the attack was devastating. All of the crew of No 2 gun perished immediately, blown to fragments. The ammunition stored in the gun-pit exploded also and the gun itself was totally destroyed with the barrel and breech burying themselves four feet in the ground thirty-five yards away. Splinters and fragments filled the air, many of them striking the Command Post where two of the Predictor crew also died, Billy Peoples being killed instantly while Bobby Hamilton died of his wounds minutes later. Jim Peoples' gun kept firing after the explosion even though he himself and several of his crew had been injured. Then after three rounds had been fired the breech-block slid back and would not return. A splinter had damaged the gravity tank and the recuperator's "cocktail" of hydraulic fluid had leaked out. With his gun out of action Jim Peoples[2] refused medical attention for himself until his wounded crew members — three Basutos — had been seen to. Bombardier Charles McCarter took over Sgt Peoples' gun but it could not be put back into action. However, its firing handle and intermediate firing lever were removed to put No 3 gun back into action. No 4 gun had also been hit and damaged but kept firing throughout the engagement putting up 111 rounds.

[2] Jim Peoples was so severely wounded that he was not expected to live. He did however survive but until his death in December 1987 still carried shrapnel in his body from that night.

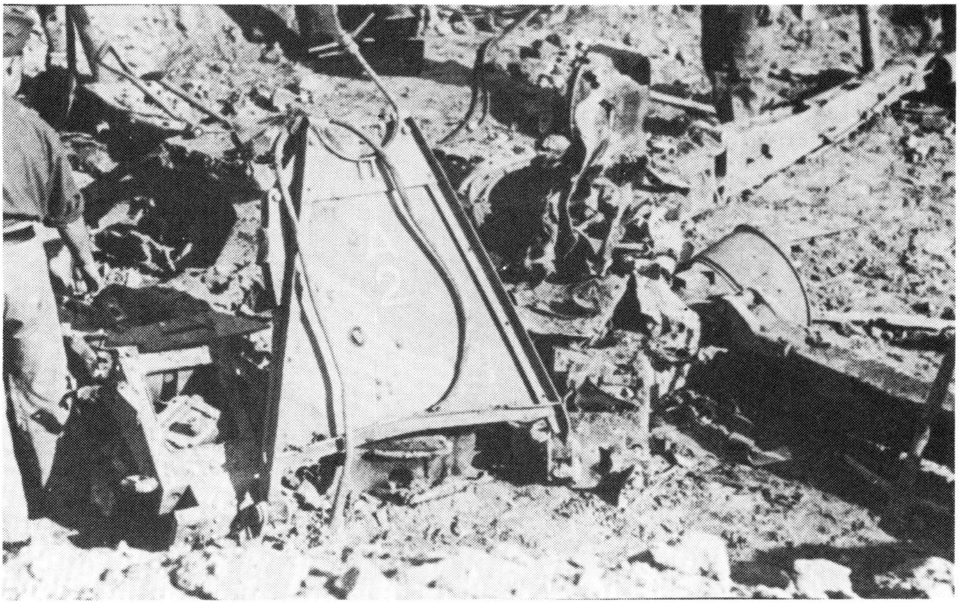

The remains of A Troop/25 Battery No 3 gun after the direct hit on the night of 21st October 1943.
(Photo: Sir Basil McFarland's Collection)

The barrel and breach of the gun were hurled 35 yards and embedded four feet into this embankment.
(Photo: Sir Basil McFarland's Collection)

Emergency services had gone into action almost immediately and those who were wounded received prompt and efficient attention with the Regimental Medical Officer, Captain Ronnie Cunningham, rendering invaluable service.

The raid lasted some forty minutes after which the Battery personnel were able to turn their attention to the horrific task of clearing up the site of the explosion and finding what might be left of their comrades. Dead were:—

Pte Selet Chele
Gnr Aubrey Evans*
L/Sgt Jim Gilmour
L/Bdr Bobby Hamilton
Pte Lesiame Hlakane
Sgt Choni Kharasi
Pte Ramaisa Khoamyane
Pte Tsoebilane Lepota
Bdr Dougie Mark
L/Cpl Solomon Mosestheli
Pte Morinanile Nena
Pte S Ntsoebe
L/Bdr Billy Peoples
Pte Azael Semane
Pte Nkhobiso Tsoasane

*"Ginger" Evans, who thought he would be safer in a gun pit.

A total of five men were wounded, Jim Peoples' condition being critical for some time.

Unknown to the members of No 1 gun crew at the time one metal splinter had passed right through the barrel of the gun pushing jagged edges into the barrel's interior. According to the ordnance experts who examined the gun next day it could not have fired after being damaged. They were dumbfounded to hear that the gun had fired three rounds in that condition! Only a miracle, they concluded, had saved that gun from destruction also as the first round fired after the splinter penetrated the barrel should have exploded in the gun. Had that happened several more names would have been added to the toll of dead and injured.

The task of clearing up and identifying the dead went on into the next day. It was a job that stamped an indelible memory into the minds of those involved. During the 22nd the remains of the dead were buried in the Military Cemetery at Volmero, where they rest in Naples to this day.

Following their tragic experience A Troop/25 Battery were visited next day by their own Battery Commander, Sir Basil McFarland, accompanied by General Rutledge (the Commander of the US Army AA Artillery) and General Matt Ridgeway (the Commander of the US 82nd Airborne Division). Sir Basil had of course also been to the devastated gun site the previous evening, immediately following the explosion, and had done much to help the morale of his men.

The entire Regiment had been in action that fateful October night. 24 Battery had noted the reluctance of the bombers to come into the range of the gun defence

"They shall grow not old . . . ". 25 HAA Battery's dead are laid to rest on October 22nd 1943 at Naples.

(Photo: Sir Basil McFarland's Collection)

and they believed that only about one-third of the intruders pressed home their attacks. Some bombs fell in the harbour area but the damage there was only slight[3]. 24 Battery's A Troop fired 407 rounds and another 357 rounds went up from B Troop.

A Troop/26 Battery had fired only 17 rounds before losing contact with the Gun Operations Room. B Troop/26 Battery however fired 333 rounds. Some bombs landed near A Troop/26 Bty but did not explode.

In spite of its sufferings A Troop/25 Battery fired 153 rounds of which 111 came from No 4 gun, which rate of fire was believed to be of record standard. B Troop/25 Battery fired 279 rounds, giving the Regiment a total of 1,456 rounds. 215 HAA Battery's two troops put up a total of 499 rounds — only 67 more than the stricken 25 Battery.

As the raiders had attacked at heights between 1,000 and 5,000 feet the light guns were also able to engage effectively. That morning in fact the existing light batteries had been relieved and the newcomers therefore had a rapid baptism of fire at Naples. Nonetheless 37, 122 and 328 LAA Batteries fired a total of 4,154 rounds.

The Germans had made extensive use of flares and some of their aircraft, including the machine which bombed 25's position, had also carried out strafing attacks using their machinegun and cannon armament. At the end of the raid however the *Luftwaffe* had lost four bombers and the destruction of two of these was credited to the guns.

Acting on the information from L/Bdr Sherrard a party of men from 25 Bty under Captain E H Babington had located and searched the house from which the light had been seen flashing before N11 had been attacked. The sole occupant of the house turned out to be an American colonel but Captain Babington forced the colonel to telephone his HQ to verify his identity! When the roof of the house was checked footprints were found in the dust but the American insisted that his Italian servants had not been on the roof and had fled as the raid began. They did not however return.

On October 20th the Regiment had come under the command of 22nd AA Brigade on leaving 12th AA Brigade. It was now no longer part of X Corps. Indeed a relief for the Regiment at Naples had been announced in the form of 59th HAA Regiment but it was to be some time before 9th Regiment left Naples. The CO of 59th HAA Regiment arrived in Naples on October 23rd to take up the post of Deputy AADC to Lt Col Aitken. His Regiment was due to follow within a week.

The *Luftwaffe* raided Naples again on the 23rd. As with the raid two nights before, it began at about 6.30 p m. Some thirty bombers were involved and they tried to outskirt the barrage rather than face it. During a period of one hour fifteen minutes all three batteries of the Regiment joined in the harbour barrage. One 'probable' was claimed by 24 Battery who fired 213 and 320 rounds from A and B

[3] Other bombs however fell in a troop concentration area claiming a number of lives including at least one Royal Artillery gunner who is buried alongside the 25 Bty men in the British Military Cemetery at Naples.

Troops respectively. The depleted 25 Bty put up 201 and 193 rounds from A and B Troops while 26 Battery's gunners were really on their mettle with 308 and 328 rounds from their A and B Troops. When the raid ended at 7.40 p m one ship in the harbour had been set on fire but no major damage had occurred. Yet again the gun defences had proved effective in deterring the efforts of the German airmen.

Lt Col Aitken was detached from the Regiment to Royal Artillery HQ, X Corps from October 26th to December 10th. In his absence Major Grant assumed command of 9th Regiment but Lt Col Miller-Jones of 59th HAA Regiment took over as AADC, Naples.

A reconnaissance Messerschmitt 210 at 27,000 feet was spotted by 26 Battery on October 29th but was out of range of the guns. However, at 9.55 a m next morning a Junkers Ju88 at 22,000 feet was fired on by A Troop/26 Battery with five rounds and A Troop/25 Battery with seven rounds.

With the imminent arrival of 59th HAA Regiment to strengthen the Naples defences it was decided to re-deploy the guns of 9th Regiment. This exercise was to involve 25 Battery moving to a single location at Marianella with their seven guns. In the case of 24 Battery, A Troop was to move from N6 position to join B Troop at N4 and thus to create an 8-gun position. Similarly, 26 Battery would concentrate in an 8-gun position, this being N3 with B Troop vacating N1. The vacated sites were to be taken over by the reinforcing batteries. However, it would be mid-November before these moves took place, and before that happened the Regiment was in action at their original sites on four occasions.

November was a particularly cold and wet month — much more akin to conditions at home in Ireland than to what the gunners had expected of Naples. The weather conditions created some problems in the completion of site-work and the provision of accommodation especially for 26 Battery. In the early part of the month however the weather did not seem to hamper the *Luftwaffe*. At 4.00 p m on November 1st a recce plane, identified as a Ju88 by 26 Battery, flew over Naples. Both 25 and 26 Batteries noted the aircraft with A Troop/25 firing 3 rounds on height control. It was however out of range to 26 Battery.

The information relayed by that recce flight was sufficient to justify a raid on Naples harbour that evening. The commander of the *Luftwaffe* in Italy, *Generalfeldmarschall* von Richthofen, a member of the same Prussian family as the "Red Baron" of twenty-five years before, had indicated to Kesselring that the best use to which his dwindling resources in aircraft could be put was to attack the Allied supply ports. Careful husbanding of his aircraft could help do more damage to the Allies than committing them piecemeal to aerial combat with the Allied air forces. Richthofen had fewer than 800 aircraft left from the nearly 1,300 available to him in July 1943. Thus it was that ports with shipping tied up were his prime targets and on November 1st there were plenty of supply ships tied up in Naples.

Luftwaffe bombers arrived over the Naples area at 6.30 p m. 25 Battery's War Diary records that once again there was no early warning. About two dozen aircraft attacked in a raid that lasted just over half-an-hour. The attackers dropped flares to mark the target area and 24 Battery's Diary notes that these were coloured green, orange and white. The Smoke Companies were brought into operation to produce a smoke-screen to obscure the harbour, and the guns fired barrage after barrage.

A number of aircraft got into the harbour area — about fifteen in all — and carried out shallow dive-bombing attacks on the anchorage. Some bombers launched torpedoes against the ships but fortunately there was no major damage. The torpedo-bombers were deflected by the intensity of the anti-aircraft fire and confused by the smokescreen. In addition the searchlights were operating and illuminated three aircraft, two of which were bracketed by fire from 24 Battery.

A Troop/24 Battery fired a total of 242 rounds while B Troop put up 350 rounds. 25 Battery fired a total of 409 rounds, some at aircraft caught in the searchlights and the others on controlled barrage fire. The greatest number of rounds was fired by 26 Battery. Their total of 815 rounds (404 from A Troop and 411 from B Troop) represented an average firing rate of 25 rounds per minute for the 32 minutes during which the Battery was firing. Not surprisingly the Battery War Diary entry for the following day notes their main task as being "replacing ammo and returning empties" — "empties" referring to the used shell-cases which littered the area around the gunpits during an action. A total of 1,819 rounds was fired by the four heavy batteries involved in that half-hour, all but 103 of them coming from the guns of 9th Regiment. Although a number of aircraft were caught in the searchlights and engaged by the guns no claims were made for aircraft destroyed or damaged.

A Troop/26 Battery at N3 position had reported a new development in German tactics — the use of "Window". This was the codename given to strips of aluminium foil dropped by attacking aircraft. Being metallic the foil showed up on radar and the effect of large quantities of small strips was to cause such a mass of "returns" on the screen that the actual position of the attackers could not be accurately calculated. This, of course, would mean that GL radar would be "blind" as would AI (Airborne Interception) radar in night-fighters. "Window" had first been used by RAF Bomber Command over Hamburg that summer. In fact its introduction into use had been delayed in case the Germans should copy the idea and use it themselves. The Naples raid of November 1st 1943 was one of the first instances of German use of "Window"; on this occasion however it "caused little interference".

Luftflotte 2's bombers came back to Naples four nights later. This attack came just before 6.00 p m, and there was ten minutes advance warning of the raid for the guns. Over twenty bombers, most of them the highly effective and versatile Ju88s, attacked over a period of 63 minutes.

Once again the Germans made liberal use of flares. Following the dropping of the first flares a number of machines made shallow dive-bombing attacks on the harbour, while others remained at between 8,000 and 10,000 feet to drop their bombs. Some aircraft in fact came so low and so close to 24 Battery's position at N6 (A Troop) that the position's machine-guns were also brought into use. The barrages effectively turned away most of the bombers but some got through and bombs fell in the docks area, starting a fire which raged for an hour before being brought under control. Whilst A Troop/24 Battery's guns were firing on one bomber their GL set crew noted the sudden disappearance of the 'target' from their screen. As there had been no interference with the radar A Troop believed this machine to have crashed after being hit by their guns.

Barrages from 24 Battery totalled 873 rounds (A Troop — 335; B Troop — 538) and their machine guns fired 345 rounds. B Troop/26 Battery claimed the destruction of two enemy aircraft and these hits, witnessed by the Command Post personnel, were later confirmed. B Troop had fired a total of 479 rounds while A Troop's total had been 298. There is no record of the raid in 25 Battery's Diary but the Regimental HQ War Diary shows a total of 2,260 heavy rounds fired. As 24 and 26 fired 1,650 of these it is probable that the remainder were fired by 25 Battery.

On November 10th the bombers struck at Naples again but on this occasion they chose to come in at 3.15 a m. The guns were ready for the dozen or so aircraft which approached the defended area. Some of the intruders turned away as the initial barrages were fired, while the remainder tried to get through. The action lasted for an hour and none of the bombers got near the harbour. B Troop/24 Battery claimed the destruction of one bomber at 3.45 a m. This machine was hit at 5,600 feet and crashed at Herculaneum. A captured crew member, the wireless-operator, confirmed that the plane had indeed been hit by heavy anti-aircraft fire. 26 Battery claimed the destruction of an intruder also but only one 'confirmed' was logged by Regimental HQ so that it is most likely that both 24 and 26 engaged the same machine.

25 Battery made no claims but their War Diary notes that this was the first occasion on which the Predictor linkage system was used, another step on the road towards automated fire control.

The Regiment's guns fired a total of 1,115 rounds during the hour of the raid. Of this the largest contribution — 452 rounds — came from the seven guns of 25 Battery while 24 Battery fired 253 rounds and the remaining 410 came from 26. There was no damage to Naples harbour.

B Troop/26 Battery fired 16 rounds at a hostile reconnaissance aircraft on November 13th. In this incident just after 1.20 p m the other guns were not involved and the aircraft turned and flew away from the defended area.

That same day work began on the preparation of a new 8-gun position at Marianella for 25 Battery, to be designated N13. Each Battery also had a visit from the Inspector of Gunnery and his assistant and was informed that the Regiment would consolidate on to three 8-gun sites. Thus by 3.00 p m on November 14th both Troops of 24 Battery were sited at N4. However, a Battery of 100th HAA Regiment moved in to N13, thus cancelling 25 Battery's move to that site. Three days later 26 Battery concentrated on N3 site with N1 being taken over by a Troop of 167 HAA Battery. The Regiment now occupied two 8-gun positions and two 4-gun positions, although one of the latter still had only three guns as no replacement had been received for the gun destroyed on October 21st. Some reinforcements were received by 25 Battery during November to bring them back up to strength.

26 Battery were busy building new accommodation and a new Battery HQ, a site for which was not identified until three days after the 8-gun position was established. They had also lost one of their Basutos, Gnr Mofosi dying in hospital on November 15th from haematomesis.

A recce plane overflying Naples on November 26th at 1.00 p m was fired at by all three batteries. The Ju88 was shot down a few minutes later by a Spitfire.

At 6.24 that evening early warning radar picked up a number of aircraft approaching Naples harbour. The guns were already on "stand-to" and when the raiders arrived the air was thick with anti-aircraft fire. At least twelve aircraft were involved in the raid and their view of Naples harbour was obscured by a smokescreen which flares failed to penetrate. The smoke also blacked out some of the gunsites so that the gunners could not see their targets. Nonetheless targets were plotted on radar and barrages fired on this information, as well as by utilising the standard barrages laid down for the defence of Naples. 9th Regiment's guns fired 1,888 rounds with 25 Battery's seven guns firing an average of just over 101 rounds each for a total of 708 rounds, followed by 26 Battery with 612 rounds and 24 with 568 rounds. No damage was done to the harbour and no claims were made by any of the gunsites.

This raid was the last occasion on which the Regiment was engaged against German aircraft at Naples. Although there were two reconnaissance over-flights in early December these were not engaged, and orders to move out of Naples were received on Christmas Eve. The foul weather which was causing so many problems to the ground troops was severely restricting air activity too.

Richthofen's diminishing resources also helped to curb the *Luftwaffe's* operations but that did not prevent them from inflicting the greatest damage to Allied shipping since Pearl Harbour in early December. This occurred at Bari on the east coast of Italy on the evening of December 2nd. Bari was chosen by Richthofen because of the concentration of shipping in the harbour and on this occasion his bombers, using "Window" again, sank or damaged many of the ships in the harbour and killed almost 100 US sailors. The civilian death toll was never accurately noted but was believed to be in the order of 700. Most of the dead were poisoned by mustard gas which had been part of the cargo of one of the American ships destroyed in the raid.

The Bari raid illustrated that the *Luftwaffe* could not yet be written off as a spent force and that the job of the AA gunners would continue to be important. Thus even during slack periods 9th Regiment would continue to train in order to maintain the high standard which it had already achieved.

CHAPTER FOUR

"VIVA VENTE, VENT'UNO!"

Even on the Salerno beachhead Italy had been completely different from the gunners' experiences of the past four years. The green grass, the flowing water, apple trees, tomato plants and above all the rain that came down in bucketfuls, just like home, all signified Europe.

And the people were different too. Genuinely pleased to welcome the Allies in spite of having been on the "other side", their greetings in Naples were almost embarassing, with men as well as women hugging and kissing the gunners as they drove in. In fact Italian civilians actually helped in the construction of gunpits in Naples.

Soon they began to make friends with Italians. Meals were shared with Italian families and rations passed on to them as well (no doubt quite illegally). Gunners were invited to special occasions in the lives of Italian families — J J Doherty attended the First Communion of a young Italian boy whose family he had befriended. Jim Dinsmore "adopted" an Italian teenager who had been orphaned and who wanted to move up north with the Regiment so that he could join the partisans and fight the "*Tedeschi*" (Germans). The would-be partisan proved to be an extremely skilled poacher and helped Dinsmore's section to supplement their rations quite nicely. Many friendships blossomed, including that between Dan Missen and a young Italian girl called Rosetta; when they married in Naples theirs was the first marriage between a British soldier and an Italian.

B Troop/25 Bty had taken over the *Stadio Littorio* where twelve partisans had been killed by the Germans in a battle just before the city of Naples had fallen. The Italians asked permission to bury their dead — some of whom were only boys — in the stadium and this was granted, with the "Derry Boys" as witnesses to the sad ceremony. And when A Troop/25 Bty suffered so heavily on October 21st the sympathy of the local community was demonstrated at the funeral of the dead gunners.

In simple ways too did the friendship show. Ninth's vehicles all carried the identification number 2021 and their Italian neighbours would greet them by calling "*Viva vente, vent'uno*" ("long live twenty, twenty-one")! From Salerno to the Arno the same warmth was shown and it is small wonder that Italy is remembered fondly by men who would prefer to forget many of the other places in which they served.

CHAPTER FIVE

ADVANCE ON ROME

The move from Naples had been planned a week before Christmas 1943 when the CO had visited the Headquarters of 66th Anti-Aircraft Brigade, under whose control the Regiment would pass, to discuss deployment of the "Ninth" at new positions in the Torre Annunciata and Castellammare areas. Both of these locations were small ports also in use for bringing in supplies to the Allied forces in Italy. When Lt Colonel Aitken together with the Battery and Troop Commanders went on a reconnaissance to select their positions on December 20th they also recce'd sites for the defence of new airfields at Vesuvius and Pompeii.

Their movement orders, received on Christmas Eve, confirmed the inclusion of the landing-grounds in the Regiment's duties and appointed Lt Colonel Aitken as AADC Torre Annunciata and Castellammare. In addition Major Jimmy Henderson,

24 Battery, and Major Sir Basil McFarland, Bt., 25 Battery, were appointed AADCs for Vesuvius and Pompeii respectively.

Christmas Day was spent relaxing, but the following day the gunners began to prepare for the move which began with Regimental HQ and B Troop/25 Battery leaving Naples on December 27th. Two days later all of the Regiment had moved out of Naples and were establishing themselves in their new positions — in which they welcomed the New Year of 1944.

On January 10th 1944 two additional gun-sites were added to the defences of Torre Annunciata and Castellammare. These were guns of an American Battery and it was the first time that "Yanks" had come under the command of 9th Regiment.

An additional duty was added to those already handled by the gunners — that of defence of ships engaged in beachlanding exercises. These exercises were directly linked to the Allies attempt to operate a "Pincer movement" on the Germans by landing a large force behind the main German defensive line.

Since the fall of Naples the Allies had slowly and painfully fought their way northwards. Italy lends itself to the defender; across the peninsula which is only 100 miles wide run a series of river valleys flowing down from the spine of the country which rises in places to some 6,000 feet. In that dismal winter of 1943/44 the roads were often snow- or mud- or water-covered. Not surprisingly the Germans made full use of both topography and weather and the freezing cold also rendered many American soldiers unfit for battle. [Not that British soldiers felt the cold less, but British officers had always carried out regular and strict checks on the condition of the feet of their soldiers. This did not happen in the US Army, with the result that hundreds of soldiers developed trench-foot simply because they did not take off their shoes for days or even weeks at a time. In many cases gangrene set in and feet as well as shoes had to be cut off. Hundreds more had to be treated for swollen feet and for a time trench-foot was causing more American casualties than were the Germans.]

As the advancing Allies crossed one river the Germans would fall back to another. Each "success" was achieved at horrific cost and German ground forces at the turn of the year outnumbered the Allies in Italy. Added to these difficulties was the withdrawal of some of the best Allied troops to prepare for the invasion of France. Out of Italy went 7th Armoured Division (the "Desert Rats"), 50th Division and 51st (Highland) Division from the British forces; from the Americans went 1st Division ("Big Red One"), 9th Infantry Division, 2nd Armoured Division and the famed 82nd Airborne. The forthcoming invasion of France also took Generals Eisenhower, Patton, Bradley and Montgomery from the Mediterranean theatre. Clark was left in command of two US Armies — 5th and 7th — while Sir Oliver Leese took over 8th Army from "Monty".

Churchill, recovering from a near-fatal illness in North Africa, now began to take an interest in Italy again. His was the mind that spawned the idea of an amphibious landing near Rome and he fought to keep sufficient landing craft in the Mediterranean to see it through. Clearly such a strategy was needed, for in the space of six weeks in November and December 1943 eight allied divisions had sustained 16,000 battle casualties and had advanced a mere seven miles!

The new plan called for the amphibious landing to be supported by an attack by Clark's 5th Army on the Rapido River. Indeed this attack would be launched before Operation "Shingle" (the amphibious landings) and would hopefully divert German troops away from the area of those landings — Anzio. Operation "Shingle" would see the rapid capture of Rome while the remainder of 5th Army would advance to join the Corps landing at Anzio. In the view of 5th Army planners it would be a straight-forward operation assured of success since "it would appear doubtful if the enemy can hold an organised defensive line through Cassino against a co-ordinated army attack".

36th (Texas) Division launched the attack on the Rapido on January 20th. Some 10,000 American soldiers advanced into the greatest disaster for their nation since Pearl Harbour in 1941. The Germans fought ferociously and in two days flung back the attack. 36th Division lost 2,000 dead, wounded or captured from two regiments alone (56% of their initial strength). 15th *Panzer Grenadier* Division opposing them suffered 243 casualties (64 of whom died) in the battle.

As the men who had trained at Castellammare went ashore at Anzio on January 22nd they did not know that half of the operation had already failed. First ashore were the men of the British 1st Division followed by the US 3rd Infantry Division. Darby's Rangers, an American Commando Unit, had gone in at Anzio port before any of the divisional landings. But the Corps Commander, General John Lucas, decided to dig in and in so doing allowed the Germans, who had provided virtually no opposition to the landings, to move troops up to the beachhead area. By the end of the month the men at Anzio were surrounded by Germans. The high hopes of a rapid advance on and capture of Rome were doomed. And Clark had acquired a new nickname; to the men of 36th Division he was now "Butcher" Clark!

At the end of January "Ninth" were well established in their new positions; Lt Colonel Aitken also had responsibility for a Light AA Regiment, an American LAA Battalion, a Smoke Company and an RAF Unit, while at Vesuvius Major Henderson of 24 Battery was also responsible for an American LAA Battalion. There had been no action against enemy aircraft. The Smoke Company at Torre Annunciata was relieved during January by an Italian army smoke unit; although now fighting the Germans the Italians were to be regarded as "co-belligerent but not as allies".

Once again the nomenclature of the Regiment's sub-units had changed, with the Troops lettered consecutively from A to F. Thus A and B Troops of 25 and 26 Batteries became C and D Troops and E and F Troops respectively. A and B Troops/ 24 Battery remained unchanged.

Thanks to the efforts of Sir Basil McFarland a Regimental Rest-Camp was established at Ravello[1] using a hotel in which Sir Basil had holidayed before the war. A rota for leave was established, allowing twenty-five personnel to go to Ravello at any one time for a five-day period. It was a popular measure as it helped immensely to

[1] The Italian Royal Family were also living at Ravello and the Gunners saw them often. The King is believed to have assumed that the soldiers in the Hotel were there to protect him.

relieve the tedium at the cold, wet gun-sites.

In addition to his role as AADC, Lt Colonel Aitken also assumed the post of Officer Commanding, Troops, in the area, in which position he became responsible for defences against possible enemy landings.

At the end of January 25 Battery was at last told that they were to be issued with another gun to replace that destroyed at Villa Nuova. Accordingly, a party under Captain Harpur went to Catania in Sicily to collect it, returning to the Battery on February 7th.

February 1944 was one of the quietest months the Regiment had experienced since June 1940. Although normal alert status had to be maintained no enemy aircraft appeared in the defended areas at all. Much time was spent on training tasks with the guns and instruments. The move to 66th AA Brigade had brought with it the need to trade-test personnel to that Brigade's standards. These were high but the men of the 'Ninth' had no difficulty in surpassing them. In 26 Battery Major Jack McVeigh had doubted if "any of the Basutos can make it". They did. By the end of February McVeigh had received notice of a transfer to the UK and 26 Battery was to receive a new Battery Commander.

The training programme continued throughout March, and the Regiment reduced its transport holding to that of a semi-mobile unit. An alarm on March 15th saw no action at any of the sites, as the bombers concentrated on Naples. St. Patrick's Day was celebrated in traditional fashion and on the following day Mount Vesuvius decided to join in the celebrations.

Vesuvius had not erupted for eighty years, and only a few days before this eruption men of the Regiment had been at the top of the mountain looking down into its crater! Now from that crater poured streams of lava while clouds of volcanic dust and ashes spewed high into the sky. Eruptions continued for more than a week, not easing until March 28th; dust continued to fall even after that date.

During those days gun sites had been covered in ash and forced to go out of action. Ash had burned through tents and the canvas covers of vehicles, while roads had become impassable. Radio communications had been blacked out by the effects of the eruption, and men from the Regiment were organised to assist in the evacuation of civilians. Visibility was at times down to one hundred feet while at other times it was normal, depending on wind conditions. The smoke from the crater was observed to rise as high as 10,000 feet[2] and at San Giuseppe cinders fell to a depth of six inches.

The city of Pompeii had been destroyed by Vesuvius centuries before, and many of its buildings had been excavated to become tourist attractions. As they watched the volcano throughout those days men who had seen the museums of Pompeii and the shapes of people frozen for ever in the attitudes in which the lava had caught them so long ago feared for their own safety. It was a fear more intense than any felt in action for, as one gunner said, they were so busy during air raids that there was hardly time to be afraid.

[2] From a 24 Bty Heightfinder reading. Another reading, by 25 Bty had earlier put the height at 20,000 feet. As some days separate both readings there is no real contradiction.

Roads in the area were cleared of lava by the end of March, at which time the Regiment was on the move yet again. 24 Battery moved to an eight-gun site at Salerno, relieving 321 Battery of 100th HAA Regiment. The airfields of Vesuvius and Pompeii were unusable due to the ash and cinders, and many aircraft had been damaged. A force of six light guns was left to defend Vesuvius airfield and 25 Battery moved again to sites in the Naples area — N14 and N15 — where it came under the operational command of 22nd AA Brigade. 26 Battery remained in the Castellammare area, although F Troop moved to a new site. Command of 26 Battery was now in the hands of Major R C Stead who had come out from England to assume this post.

At Salerno Major Henderson became AADC and took over operational command of 347 LAA Battery and a smoke company. The smoke was withdrawn however early in April.

Although 24 Battery was engaged only in practice-shooting during April the other batteries were both in action against enemy aircraft. On April 16th at Naples 25 Battery fired 96 rounds in six harbour barrages from N14 in a raid that began at 4.40 a m. Five nights later at 9.46 p m the same site fired 16 rounds at what was believed to be a reconnaissance aircraft. Six or more aircraft entered the "Inner Artillery Zone" at 11.00 p m on April 22nd but identified themselves as "friendly" after 14 rounds had been fired at them. Similar intrusions by Allied aircraft into the IAZ occurred on the last two nights of the month.

A massive attack took place on Naples on April 24th with some fifty bombers approaching the harbour from various directions. "Window" was dropped but 25 Battery fired a total of 458 rounds in harbour barrages on height control and on unseen target procedure. 330 of those rounds were fired by N14 and although one aircraft was illuminated no claims were made. 26 Battery was also in action during this raid, firing a total of 103 rounds. Three aircraft were engaged but no claims made. "Window" was noted by 26 Battery also, but it created no problems in tracking the planes.

All three batteries saw action during May. At Salerno 24 Battery fired at enemy aircraft flying over the area. Barrage fire was used as the radar was unable to track the targets because of "clutter". The aircraft flew off to the west as the guns fired. One bomb was dropped but no casualties or damage occurred. A total of 86 rounds had been fired. That action took place on May 14th, a day after 26 Battery had been in action against enemy bombers over the Bay of Naples from 3.34 a m to 4.00 a m. E Troop had fired 70 rounds at an enemy aircraft dropping a torpedo and F Troop had fired 21 rounds. In addition 43 rounds had been fired from the new American ·50 calibre machine-guns which had been issued to the Regiment.

25 Battery's guns burst into action on May 15th when, at 4.27 a m, more than twenty aircraft attacked Naples. Most approached from the north and the Battery fired a total of 291 rounds all but 36 of which came from C Troop. No hits were claimed.

It was during this period of activity that 9th Regiment said goodbye to its distinguished CO as Lt Colonel Aitken left to join 11th Field Regiment. During his time

A warscape on the road to Anzio as a Don R leads the way through the shattered countryside around Cassino.

(Photo: collection of Sir Basil McFarland)

with them he had earned the respect and admiration of the men of 9th Regiment. In his turn he never forgot the Regiment, and when a memorial was unveiled to them in St. Columb's Cathedral in 1962 he travelled to the Regiment's home city to unveil it. Of him Sir Basil McFarland wrote — "Others got the kudos but he did the work. A shy man, a great disciplinarian, firm and fair with the men, they got to know and love him during his 2½ years with the Regiment".

With German air activity reducing even more in Italy there was an opportunity for some of the anti-aircraft units to operate as field artillery. The 'Ninth' had already practised this role and it was nothing new to them when they began to train as field gunners again. One of 24 Battery's Troops moved to San Angelo as Depot Troop of the firing range there and played host to all three batteries as they each sent their Troop Commanders, Observation Post Parties, Command Post and Assistant Command Post Officers for training in field shooting.

The training programme ended on May 23rd. On the same day Operation "Buffalo" saw US and British troops break out from Anzio. Five days earlier Cassino had finally fallen; Fifth and Eighth Armies had broken the Gustav Line on May 16th; at Terracina on May 25th the Anzio forces linked up with Fifth Army. With Rome now in sight 9th Regiment was warned to be ready to move with 66th AA Brigade on May 28th. They were relieved by 71st HAA Regiment and moved into a concentration area at Torre del Greco on May 30th. From there they would move north to Anzio.

CHAPTER SIX

THE ETERNAL CITY

On June 4th 1944 Rome greeted its latest conquerors — the soldiers of Fifth Army. General Mark Clark had achieved what Hannibal had failed to do and had captured the Eternal City. In Hannibal's case however the city had been defended while Clark was able to enter a city which both sides had agreed to recognise as "open" following the pleas of Pope Pius XII. Clark's anxiety to enter Rome before Eighth Army and thus make the liberation an American affair saw him quite deliberately misconstrue the plans for the Allied offensive and then defy Alexander. Fifth Army entered Rome with no problems because Clark had failed to close the pincer with which Fifth and Eighth Armies were intended to envelop the retreating Germans. The "liberation" of Rome and Clark's newspaper headlines were to mean that the campaign in Italy would be prolonged even more as the Germans regrouped and prepared for a new defensive to the north of the capital.

Clark's moment of glory was short-lived. On June 6th 1944, under the command of General Montgomery, Allied ground forces landed in Normandy. D-Day was to become the date to remember from that summer month of 1944 and the purgatory of Italy was to become a sideshow.

As the Allies fought to create positions from which a breakout from the Normandy beachhead could be launched Ninth Regiment was moving into position at Anzio. RHQ, 25 and 26 Batteries preceded 24 Battery which had been detached for some days to 62nd AA Brigade to defend bridges across the Garigliano River. Just before 24 Battery had left Torre del Greco to join 62nd Brigade they had been rejoined by B Troop's radar and its operators (who had been on the island of Ischia for seven weeks under RAF control to provide early-warning of enemy air activity and who returned with the highest of praise from the RAF). By the evening of June 6th all of the Regiment's guns were ready for action at Anzio, having spent all night and the early part of the morning travelling.

There were alerts on each of their first two nights at Anzio, ensuring that there was little opportunity for rest. On neither occasion did the heavy guns fire although the light AA guns opened up both times. However on their third night at Anzio the *Luftwaffe* launched a determined attack employing more than twenty aircraft. The German planes came in at heights from 12,000 down to 500 feet and in the course of the raid, which lasted for one and a quarter hours, at least eighty bombs were dropped. Most of these fell ineffectively out to sea, the defences having discouraged the airmen from coming too close, but some did land in the harbour area, with an ammunition dump being hit and two fires started.

"Window" was once again used by the attackers but this did not prevent the gunners from putting up effective fire. All six troops were in action throughout the engagement with the highest number of rounds being fired by E Troop/26 Battery. For a total expenditure of 235 rounds they claimed one enemy aircraft shot down. 25 Battery's guns had fired 425 rounds (D Troop 224; C Troop 201) while F Troop/26 had put up 89 rounds. B Troop/24 Battery fired 145 rounds, while a further 82 rounds came from the guns of A Troop.

Although the fires in the ammunition dump blazed until morning and there were many explosions, no damage occurred to shipping in Anzio harbour. This was largely due to the efficiency of the heavy guns which had forced the pathfinder aircraft to drop their marker flares wide of the target. The aircraft claimed by E Troop/26 Battery was later credited as a "probable" although it appears really to have been a "destroyed".

As the tired, sweating gunners stood down with the "all clear" and waited for any possible further attacks before dawn they were not to know that they had fought their last major anti-aircraft engagement! The *Luftwaffe* in Italy was a much depleted force and, although it could not be regarded as beaten, its efforts in future would be but shadows of what had gone before. Because of this the Regiment would soon find itself cast, again, in the role of field artillery.

Relief at Anzio was announced on June 21st and five days later 88th HAA Regiment took over most of the Anzio positions. On June 27th the Regiment moved up to Civitavecchia to the north-west of Rome where they took over positions from US anti-aircraft units. Since the morning after the raid at Anzio the Regiment had been under the temporary command of Major Sir Basil McFarland, Bart. as Major E E Grant, who had assumed command on Lt Colonel Aitken's departure, had been admitted to hospital. As the senior of the three battery commanders, command of the Regiment devolved on Sir Basil who filled this role until June 30th when a new CO, Lt Colonel R C Mayhew, arrived. Command of 25 Battery had in the same period been in the capable hands of Captain E H "Sonny" Babington.

Hardly were they settled in Civitavecchia than they received a warning order to prepare for a move up to Leghorn (Livorno), a city which was still in enemy hands. At Civitavecchia there were only two engagements, the first of which was against two aircraft on the night of their arrival. In this action 42 rounds were fired by 24 Battery, 51 rounds by 25 Battery and 38 rounds by 26 Battery. The second action was on July 2nd when F Troop/26 Battery fired 22 rounds at an unidentified aircraft which entered the Inner Artillery Zone. In addition to these actions there were alerts on virtually every night of their time at Civitavecchia.

Batteries from both 71st and 88th HAA Regiments took over Ninth's positions at Civitavecchia on July 3rd and 4th and the Regiment moved into "laager" at Campiglia awaiting the capture of Leghorn. A tactical HQ had been established at Cecine but this was pulled back to Campiglia on July 11th. On that day also 66th AA Brigade took over responsibility for anti-aircraft defences at San Stefano and Orbetello, and early on the morning of July 12th 26 Battery moved into two 4-gun positions at San Stefano. There they came under the operational command of 12th LAA Regiment.

For the remainder of July, 26 Battery would endure a series of night alerts, none of which would lead to any action. Added to these were a number of daytime alerts all of which combined to keep the gunners on their toes. On July 21st they had a practice-shoot at night, firing at star-shells which had been fired by Royal Navy ships — one of which a group from the Battery had visited earlier that day. Although the light guns fired on July 28th and fires could be seen from Biombino on the 26th, as well as gun flashes from Civitavecchia on the 31st, the men of 26 Battery had no opportunity to go into action themselves.

August had a similar pattern for 26 Battery although the number of alerts reduced to only six for the month. Heavy air activity was noted on the nights of August 4th and 15th but in both instances this was by Allied aircraft carrying out strikes on the enemy. On August 5 the Battery had a visit from the Brigade Commander and the Brigadier — REME during which two practice-barrages were fired by each Troop.

The night of August 9th/10th gave a lot of excitement, but it was provided by the elements as the area was hit by heavy electrical storms. Heavy rain was accompanied by lightning which struck and set on fire most of the barrage balloons in the area. Several fires were also caused by lightning striking houses. It is unlikely that there was any connection between this storm and the fact that the Padre had visited the Battery to hold a Service on the day before the storm!

On August 28th the Battery's Basutos were transferred to 55th HAA Regiment while the Battery was relieved the following day by 398 Battery/79th HAA Regiment. 26 were ordered to move to a concentration area near Naples and Major Stead was posted to 153 HAA Battery, being replaced by Major Allen Welch, a Derryman and an original officer of the Regiment.

During its time at Anzio 9th Regiment had been invited to send men to join parties which were being received in audiences by the Pope. There was no shortage of volunteers to visit the Vatican and although the army had stated that the men for these audiences should be mainly Catholics those who went included quite a few whose religious allegiance was far removed from the Church of Rome! There was more than one prominent Orangeman who knelt to receive the blessing of Pius XII although few repeated the achievement of a Warrant Officer of the London Irish who wore his Orange sash to the Vatican. There can be very few sashes which have been blessed by a Pope!

Souvenirs of the visit to the Vatican were provided by the Pope in the form of a rosary for each soldier attending. These were prized by Catholic and Protestant alike and although many of the latter parted with theirs after the war they ensured that they found "good Catholic homes" for them.

One of the first into Rome was Willie Hunter, then a sergeant in 24 Battery. He recalls strolling into the Vatican with a sergeant from a Guards Regiment. As they admired and discussed the decor of the interior of St. Peter's Basilica Hunter felt a hand touch his elbow and heard a voice saying "You're a Derryman, aren't you?" He turned to meet a priest who introduced himself as Father McDaid of the Irish College, a native of Derry! There followed a long conversation about Derry and Willie Hunter returned to his Battery with a gift of wine and a rosary. Father McDaid (later Monsignor McDaid) was incidentally to become a very good friend of Sir Basil McFarland.

Men of the Irish Brigade were received in audience by the Pope on June 12th and there was more than one wag who suggested that July 12th would have been a better date. For the Derrymen however there could have been no date better for their Papal audience than the one they received — June 9th, St. Columba's Day. No doubt it was a coincidence that they got to see and meet the Pope on the feast day of their city's patron-saint, but it was a very happy coincidence nonetheless.

AUTHOR'S NOTE

GENERAL MARK CLARK

Mark Clark is now one of the lesser-known Allied commanders of the Second World War in spite of his penchant for publicity. This eagerness for personal image promotion made him most disliked by many of his colleagues. These included General George Patton who described him as always having his mind on Clark and of having large numbers of pictures taken "for the glory of FDR and for Clark where he could get a chance". Patton, no shrinking violet himself, found Clark's behaviour "disgusting". ("Clark" by Martin Blumenson; Cape, London 1985.)

His ability as a commander was called into question at Salerno when he was prepared to withdraw the American VI Corps from the beachhead and put it ashore again in the British sector. Quite apart from the physical difficulties of such an operation he does not seem to have considered the adverse effect on morale which this would have had. However his superior, General Alexander, vetoed the plan as it could only end in disaster principally because of the morale aspect. In "Alex", (Pan, 1973) Nigel Nicolson records Alexander as telling Clark that "We stay here and we die here. We do not retreat". Later in the campaign Clark disobeyed Alexander's orders in the advance on Rome and although he (Clark) had the distinction of liberating Rome he also allowed a large part of the German army to escape, thus prolonging the war in

Italy. In "Rome 1944" Raleigh Trevelyan relates an interview after the war which journalist Sidney Mathews had with Clark on the fall of Rome; Clark told Mathews that he had threatened to open fire on 8th Army should any of its units approach Rome and prevent 5th Army from gaining the glory of being the liberators!

Of course the disaster on the Rapido was a point on which many American soldiers felt that Clark should have been called to task. In fact some officers of 36th Texas Division forced an official investigation after the war but Clark, who had been decorated for gallantry and been promoted to four star general, was exonerated.

On the other side of the argument Clark has been assessed as a good leader, especially of a multinational army, and certainly Churchill was impressed with him, calling him the "American Eagle". In his memoirs Field Marshal Montgomery gives the credit for the breakout from Salerno to Clark so that, at least in this instance, Monty had a good regard for his counterpart. Certainly those men of 9th (Londonderry) HAA Regiment who met Clark were impressed by him and there is no doubt that he had a good personality.

The reader who wishes to study this enigmatic general in greater depth is recommended to read "Clark" by Blumenson and "The Long March on Rome" by Charles Whiting (see Bibliography).

CHAPTER SEVEN

TO THE FIELD — THE FALL OF LIVORNO AND THE BATTLE OF THE ARNO

The fall of Rome did not mark the end of the Italian campaign. Kesselring had already established further defensive lines across Italy north of the capital, and once again Alexander's armies would have to break those fortifications. The ultimate German line across Italy was the Gothic line from the Gulf of Genoa across the Appenines to Pesaro on the east coast. Along this line Kesselring planned to hold the Allies during the winter of 1943/44. In this he would be aided by the Allied High Command, who withdrew many of Alexander's divisions for use in Normandy and thereby prevented any swift thrust into Northern Italy across the plain of Lombardy and into Austria — the original objective of the entire campaign.

Forward of the Gothic Line was the Albert Line, stretching from near Orbetello to Lake Trasimene and thence to Gubbio. Withdrawal from the Albert Line was to be made by a series of easily defended points in the Tuscan hills and along the Arno River. And so as Eighth and Fifth Armies advanced northward yet again the "Old Hares"[1] of Kesselring's armies awaited them.

Ninth Regiment had remained at Campiglia since their arrival there on July 5th. Mark Clark's Fifth Army were pounding at the city of Livorno (Leghorn) after having broken the western flank of the Albert Line. German resistance was however particularly stiff, and thus it was that at 2.00 p m on July 12th the CO of the Ninth was told to move two Troops up to Resignano to be employed as field artillery. 26 Battery

[1] "Old Hare" — a seasoned soldier, the equivalent of the British "Old Sweat".

had already deployed to San Stefano in an ack-ack role and so it fell to 24 and 25 — who each had field artillery experience — to provide the Troops. The following day the guns prepared to move up to sites which had been chosen during the day by recce parties from RHQ, 24 and 25 Batteries. In addition to the guns of B Troop/24 Battery and C Troop/25 Battery there would be a further eight guns from 55th HAA Regiment.

Enemy artillery activity meant that the guns had to move up under cover of darkness — a nerve-wracking experience for all concerned, especially as the road was particularly bad and at one point would require winching the guns. Observation Post parties were sent forward and by 10.30 a m on July 14th the two Troops were in position.

The guns were "shot in" on targets in the Monte Nero area and the shooting on each of five targets was reported as accurate. "Shooting in" meant that the guns engaged a specific target and when the fire was observed to fall on target the relative firing data was recorded for future use. The target was then described as "registered" and given a code so that when attacking troops required support the code was given for the target they wanted to be engaged and the guns could drop shells on it immediately.

Later that day a concentrated shoot was called for on German artillery positions at Monte Nero. The two Troops fired 38 rounds each of high-explosive ammunition using Fuze 208 which created an air-burst, each shell spraying over 300 "bullets" down on its target. This fire was put down in five salvoes and had the effect of silencing the German fieldguns.

The tactical reason for the 3·7s being moved to this position and deployed in a field role was to support the US 34th Division and in particular 804th Tank Destroyer Battalion which was advancing on Leghorn. As a result of this there was a close liaison with the American Artillery and with 804th TD Bn which was to increase in the next few days.

During the morning of July 15th the OPs of both 24 and 25 Batteries moved forward along with 804th TD Bn and a Recce Troop from 34th Division. In addition the remaining guns of 55th HAA Regiment were to be brought up and that Regiment would deploy away from the site it shared with 9th Regiment. A Troop/24 Battery and D Troop/25 Battery were ordered to join their comrades as field gunners, with all of 25 Battery's guns remaining at the existing position and 24 Battery occupying a fresh position.

Throughout the day the guns were in action bringing down fire on targets requested by the advancing Americans. One of these was a tower in the Monte Nero area which was being used as an OP by the Germans. Airbursts in the area and several hits on the tower caused its rapid evacuation by the Germans. During the afternoon 804th TD Bn entered Quercianella almost seven miles to the north of 24 and 25 Batteries' guns at Castiglioncello but well within the effective range of the 3·7 when used as a field-gun. The OPs for both batteries were also moving up and that for 25 Battery came under heavy shelling at Quercianella at 8.30 p m.

This was part of a German response to heavy shelling on their positions on Hill

210. One of the shells landed close to 25's wireless truck and seriously wounded BSM Walter Wray[2], who was evacuated to an American field hospital for surgery on his injuries. The wireless truck was damaged but the wireless operator continued to transmit information. In response to this the guns of 25 Battery poured down a torrent of fire on the enemy guns. A total of 330 rounds were expended to airburst over the enemy positions. The German 88 mm and 105 mm guns fell silent.

Throughout the night the guns of both batteries kept up a pattern of harassing fire at irregular intervals on the enemy's positions. By 5.00 a m on July 16th the eight guns had discharged over 500 rounds in this manner.

The wounding of Walter Wray and the knocking-out of the German OP had emphasised the importance of forward observation and indeed the danger faced by personnel engaged on such a task. It was the role of an OP[3] to observe enemy activity in order to identify the location of enemy units which were a particular danger and to call down fire on these from the artillery. Forward Observation Officers could therefore be called on to move up with infantry or armoured units to front line positions where, if identified positively by the enemy, they could themselves become prime targets. As the soldiers manning an OP needed to have a good view of enemy positions this meant choosing vantage points such as hills or tall buildings which would of course present themselves as good targets to the opposition.

The job of the Forward Observation Officer was made slightly easier by the use of aircraft. Slow flying machines such as Austers or Cessnas were produced as Air Observation Posts (AOPs) and one such aircraft operated as an AOP for 24 and 25 Batteries during the later stages of the advance on Leghorn.

Throughout the morning of July 16th concentrated fire was laid down in support of the Americans advancing on Monte Nero and at 1.18 p m all of the guns fired on what was described as a "suspicious target". From the volume of smoke and dust rising from the area after it had been shelled it was believed that the target had been an ammunition or fuel dump.

Air observation became available that evening through the US 125th Gun Battalion, and B Troop/24 Battery put down fire on several targets on the southern outskirts of Leghorn. A Troop/24 Battery had also come into action, and between them the two Troops fired 481 rounds at various targets during that evening and night. 25 Battery's guns fired, at the request of 804th TD Battalion, at enemy infantry concentrations, anti-tank guns and at various road intersections. The AOP pilot especially commented on the speed and accuracy with which both batteries engaged their targets and the US field artillery units also brought the effectiveness of the British anti-aircraft gunners to the attention of Fifth Army HQ. During the night C Troop/25 Battery continued harassing fire in support of 804th TD Bn, as their

[2] Walter Wray was arguably one of the most popular men ever to have been a Warrant Officer. He recovered from his wounds sufficiently to return home with the Regiment in September 1944 and in 1945 he became RSM of 9th Regiment. His close friend, Jim Canning succeeded him as BSM of 25 Bty.

[3] Billy Smyth was a regular member of OP parties. He found the job fascinating as he had a natural aptitude for the mathematics involved. When 9th returned home he was transferred to a Field regiment and served in NW Europe.

patrols had encountered further enemy resistance. By dawn the four guns had fired a total of 192 rounds.

There was no rest for the gunners. After firing throughout the night they were again called on by the US 34th Division artillery to put down concentrations of fire during the morning. At 11.15 a m a concentration by all sixteen guns of 24 Battery and 25 Battery broke up a formation of German tanks and infantry. Later that day however the two Batteries began to run short of the very effective Fuze 208 ammunition, and this was then restricted to use only to repel a "determined counter-attack". Fuze 199 shells only were to be fired on the basis of fifty rounds per Troop per day until fresh supplies became available. Fortunately that evening lorries of 45 LAA Regiment were able to bring up further stocks of ammunition, enabling 25 Battery to fire 289 rounds on enemy guns in the Monte Nero area. The ammunition situation was not however good enough to allow for a programme of harassing fire throughout the night.

Early in the morning of July 18th 25 Battery silenced enemy mortars with 12 rounds. That afternoon C Troop moved up to a new position while A Troop/24 Battery knocked out more mortars with 24 rounds of Fuze 208 ammunition.

At 8.00 a m next morning came the news that Leghorn had fallen. 24 Battery's guns were ordered to concentrate on A Troop's position while C Troop/25 Battery moved back to rejoin D Troop.

Orders were now issued for 24 and 25 Batteries to deploy for anti-aircraft defence at Leghorn. On their way to recce the positions allocated to them at Leghorn Major Jimmy Henderson and Captain Paddy Laird of 24 Battery had a graphic description of the effects of their gunfire from inhabitants of Monte Nero. It had been "terrifying" said the Italians, while one partisan told them that ten of the *tedeschi* (Germans) had been killed by a single airburst. These rounds had been the most frightening of all and had affected the morale even of Kesselring's "Old Hares".

On reaching Leghorn it was found impossible to recce the positions as the area north of the canal was still in German hands and only one site could be selected.

24 Battery left their concentration area at Rossignano-Solvay late on the night of July 22nd to move into Leghorn. On the way part of the road collapsed under an ammunition lorry borrowed from 25 Battery. The lorry overturned and three men were trapped under three tons of ammunition. Private Sefako Motuba died later of his injuries and another African soldier was seriously injured. Before they reached Leghorn they had another, less serious mishap, when a small bridge collapsed under the weight of the convoy. The bridge was repaired and by 2.30 p m the Battery's guns and radar were ready for action.

A few days before they had received a new GL No 3 Mk II but, as Leghorn was still being shelled, this was left at Rossignano.

25 Battery moved into Leghorn on July 24th and that evening were warned that C Troop was to return to its field artillery role. Next day the German artillery shelled Leghorn yet again with several shells landing in 24 Battery's area. There were no casualties although one round burst close to Battery HQ.

C Troop/25 Battery returned to its field role on July 26th moving to a new position at 7.30 p m. Less than three hours later at 10.12 p m two aircraft attacked and bombed the harbour. By the time 24 Battery received permission to fire the planes were away and only a single round was fired by A Troop. Next day 66th AA Brigade gave permission to engage immediately any aircraft entering the Inner Artillery Zone (IAZ); that evening B Troop fired 13 rounds at an unidentified aircraft.

Leghorn had quietened down sufficiently for 24 Battery to bring their new radar into operation there on July 30th.

In the meantime 25 Battery had returned to field-firing. From their new position between Leghorn and the Arno River C Troop had fired 74 rounds during the evening of July 27th. A further 53 rounds had been fired by D Troop from their AA position in Leghorn! At 11.04 p m both Troops fired a total of 25 rounds at an unknown aircraft.

Next day C Troop engaged a German mobile gun and knocked it out while two hours later, at 9.00 p m, they dispersed an infantry concentration with a barrage of 116 rounds. Both Troops of 25 were in action on July 29th with C engaging a truck and infantry personnel at 7.00 a m while one gun of D Troop had fired 20 rounds of harassing fire at irregular intervals. That afternoon Lieutenant Teague of 25 Battery reported to an OP at Marina di Pisa to pass orders and information to the Gun Operations Room. A total of 16 airbursts were fired at enemy troops in the hope that a concealed rocket gun would disclose its position by firing but the crew did not rise to the bait.

Between 10.30 p m and midnight next night one gun of C Troop fired 42 rounds of harassing fire, while two of D Troop's guns fired a total of 54 rounds at ten targets varying from 15,200 to 17,600 yards distance from the guns. As the last round fell on target 5 a large explosion occurred followed by a fierce fire for some time punctuated by further explosions, suggesting that an ammunition dump had been hit.

At 5.45 a m on July 31st C Troop fired 12 rounds at a group of enemy troops several miles away. The troops scattered and later a further 4 rounds were fired, air-bursting over woods in which enemy personnel were dispersed.

Fifth Army now held the southern bank of the Arno River from Empoli westward to Pisa, while Eighth Army was poised to take Florence. General Alexander's plans to break through the German lines and drive to Austria, a strategy enthusiastically supported by Churchill, were however thwarted by the American insistence on going ahead with the landings in southern France — Operations "Anvil" and "Dragoon". Such landings had originally been planned to assist the D-Day Operations in Normandy by ensuring that German troops based in the south of France would not be withdrawn to support the forces on the Channel coast. By the end of July it was much too late to claim such an objective for "Anvil" and "Dragoon", and Churchill argued that these should be cancelled and the major effort given to Alexander's thrust for the Gothic Line. However it was decided to proceed with the landings in southern France and during July almost 100,000 men were withdrawn from Fifth Army for the French operation. "Anvil" and "Dragoon" were to be undertaken by allied forces constituted as the Seventh US Army, command of which passed to General Patch, and which included the unfortunate 36th Division. Free French forces formed a large part of Seventh Army although they later became First French Army. Air forces were also withdrawn from Italy for this new invasion although the *Luftwaffe* strength in the Mediterranean area was now less than two hundred aircraft.

With a vastly weaker Fifth Army Clark now prepared for the next phase of operations. Alexander planned to attack with Eighth Army on the eastern side of Italy and thus withdrew two corps eastward leaving XIII Corps to come under Fifth Army command. The two corps which moved to the east concentrated with the Polish Corps near Pergola in preparation for an attack along the Adriatic Coast. Such an attack was intended to draw German troops from the centre of the Gothic Line through which Clark would launch Fifth Army towards Bologna and Imola. Success in this operation would then allow Alexander's Fifteenth Army Group to break into the plain of Lombardy and strike through to Vienna achieving Churchill's ambition of beating the Russians to Austria.

However Alexander's entire command held twenty-three divisions against which Kesselring had twenty-six German divisions, still in good fighting shape, and two Italian fascist divisions. Although these were weaker than the Allied units they had the advantage of being dug-in with easily defended territory where the Allies' superiority in armour would be of much reduced value. The Allied Operation was given the code-name "Olive" and was planned for August 15th to coincide with "Anvil" and "Dragoon". In the event "Olive" was postponed by ten days and although initially successful failed to achieve the breakthrough into Lombardy.

During the build-up to "Olive" 25 Battery continued to operate in a field role with the bulk of this work falling on C Troop which had moved up to the Arno. D Troop, although occupying an anti-aircraft site, was also used for field work. 24 Battery remained in its AA sites at Leghorn where on August 18th they provided soldiers for a guard of honour when Prime Minister Winston Churchill visited Leghorn; Major Jimmy Henderson commanded that guard.

On August 22nd B Troop were ordered forward as field gunners to a site 2,000 yards south of Pisa and the German lines. It was described as being "decidedly too close to the enemy FDLs (Forward Defence Lines) to be a health resort" and they had to move up under cover of darkness on a road on which the German artillery had "shot in".

B Troop fired 32 rounds harassing fire in support of the US IV Corps on the evening of August 23rd with a similar quantity the following evening. During the night of August 25/26th, when Operation "Olive" was launched, they fired a further 20 rounds. Their field of fire included the Leaning Tower of Pisa and the guns were ordered not to fire within an area 5° on either side of this historic building.

Before B Troop were withdrawn they heard the news of their imminent return home, the first man to hear it being Tommy McCready. Nonetheless no-one rejoiced until they had cleared the road down from Pisa safely; then the singing started! B Troop/24 Battery returned to Leghorn on August 28th where the Battery was relieved of the AA defence of Leghorn port by 246 HAA Battery of 79th HAA Regiment.

C Troop/25 Battery were in action every day from August 1st to August 26th with support from D Troop on numerous occasions. In the first seven days alone C

Troop fired 305 rounds at a variety of targets which ranged from just under 6,000 yards to almost 20,000 yards. Included in the targets were enemy tanks, artillery formations, groups of infantry, mortars, roads and strongpoints. In almost every case the OPs reported that the shelling had been successful. On one night alone C Troop put three enemy artillery batteries out of action in thirty minutes! That occurred on August 4th and on the same date D Troop/25 Battery had also silenced a hostile battery.

A convoy was broken up on August 5th while the following day C Troop demolished houses being used as an OP, smashed a mortar battery and cascaded airbursts over a group of German infantry seen in a field. More mortars were put out of action on August 7th. In the same period D Troop had fired 36 rounds all of which were recorded as effective.

The Germans had not taken all this punishment lightly but they found that they were unable to bring effective counter-fire to bear on the 25 Battery guns. They did however shell the Battery's OPs a number of times but without success, although some shells did burst close to an OP on one occasion.

The OPs continued to call down fire on selected targets and the guns were used to engage still more mortars, artillery, trucks and tanks as well as to make life extremely unpleasant for the German infantry with the fearsome airbursts. 76 rounds were fired by C Troop on August 8th knocking out two groups of mortars, destroying several vehicles and scattering groups of infantry. Mortars and guns were engaged the next day as well as troops. On three occasions mortars were silenced by eight airbursts each time; a further 20 airbursts put an artillery battery out of action and a group of infantry were scattered by 23 airbursts.

And so it continued day after day. 400 rounds were fired by C Troop between August 11th and 18th when ammunition was again rationed. In the same period D Troop fired only 10 rounds harassing fire. The fire continued to be effective and again to spell the end for many German gunners and mortar operators. On August 13th enemy mortars had engaged the Troop's OP but a salvo of 16 airbursts rapidly knocked them out. Bridges, road junctions and crossroads were included in the targets for August 16th and next day the infantry called for urgent fire on enemy mortars over a 500 yards front. An initial salvo of 24 rounds airburst was followed by a request for more support. To this request C Troop replied with 48 rounds of airburst and a further request for suppressive fire on an additional target was complied with and 28 rounds were fired. All of the mortars were silenced and a house was demolished. The OP reported enemy personnel fleeing from the destructive effects of the gun fire and it was believed that heavy casualties had been inflicted on the Germans.

Next evening at 6.00 p m German artillery shelled and hit C Troop's OP. It was estimated that 13 or 14 rounds fell in the area but no casualties were caused, recalls 25's Battery Captain, "Sonny" Babington, who was in command of the OP at the time.

While C Troop continued its firing programme on the 19th, firing a total of 32 rounds at two groups of mortars, D Troop were ordered to move up to join their comrades on the night of August 20th. The move was postponed although that day both

Four of 25 Battery's guns are actually concealed in this photograph while firing as field artillery on the River Arno near Pisa.

(Photo: Sir Basil McFarland's Collection)

Farewell to Italy! 25 HAA Battery drawn up on parade — an unusual experience for them! — to be inspected by General (later Field-Marshal) Alexander before they left Livorno to travel south to Naples and the boat for home.

(Photo: Sir Basil McFarland's Collection)

C and D Troops engaged the same target with 8 rounds each. C fired a further 8 rounds at mortars, 32 at guns and during the evening another 16 rounds at mortars. All of these engagements were described as effective and the four guns continued to fire throughout the night putting down a total of 40 rounds harassing fire on targets at ranges from 11,650 to 13,800 yards.

On the morning of August 21st both C and D Troops fired 3 rounds each and that evening C Troop broke up with 57 rounds of airburst an infantry formation which was assembling in a wood. A final 8 rounds of airburst were fired later that evening by C Troop at enemy mortars. D Troop's move to forward positions was cancelled and their place taken by B Troop/24 Battery who moved up on the 23rd and remained for three days.

Early in the morning of August 22nd C Troop silenced enemy mortars with 16 airbursts. That afternoon they engaged two German self-propelled guns at 7,650 yards. A total of 8 airbursts put the guns out of action, one being set on fire.

Guns and mortars were the targets on eleven occasions over the next three days. Each time the results were effective and a total of 159 rounds was used. A further six rounds of airburst were fired at ground troops on the morning of August 24th. The gunners continued to fire at intervals throughout the night of August 24/25th ranging in on a railway and a road junction. Ten airbursts were fired at these targets and later that day a further thirteen airbursts were aimed at troops.

On the afternoon of August 26th the advance party of 79th HAA Regiment which was to relieve 25 Battery arrived at Leghorn. C Troop fired 8 rounds at each of two mortar targets that same day before pulling back to Leghorn. The Commander-in-Chief, Anti-Aircraft, Italy, visited them to compliment them on their work on the Arno and to inform them that all of the Irish personnel and those others with 4½ years overseas service would be going home.

Before they departed they went on show for a VIP for the last time in Italy when the AA C-in-C, Major-General Hornby returned with no less a person than General Sir Harold Alexander, the Allied Supremo in Italy. As "Alex" was an Irishman they felt he was worth pleasing and the guns were cleaned, helmets painted, belts and anklets scrubbed and a parade which would have pleased a Guards Drill Sergeant followed. ("Alex" was of course an Irish Guards Officer.) After Alex had left Sir Basil gave the order "C Troop, Two Five Heavy Ack Ack Battery, cease firing" and they prepared to handover as D Troop had already done.

Both 24 and 25 Batteries left Leghorn with Regimental HQ on August 29th and travelled via Rome, where they handed in their guns, to Naples. They had bade farewell to their Basutos and to many of their non-Irish comrades who would be transferred to other units. With a mixture of elation and regret they travelled south where they awaited THE BOAT at Afragola with 26 Battery.

On September 4th the Regiment embarked on the *Orantes* for the homeward voyage. Their convoy sailed from Naples two days later, passed through the Straits of Gibraltar on September 9th and arrived off Liverpool five days later. They berthed at Liverpool on the 17th and that evening travelled by train across England to Skegness, where they arrived in the early hours of the morning to be billeted awaiting the finalisation of arrangements for them to go home on leave.

PART FIVE

FAREWELL TO FOREIGN FIELDS

CHAPTER ONE

WELCOME HOME!

Across the waters of the River Foyle the bells of St. Columb's Cathedral pealed out a greeting on the afternoon of September 25th 1944 as the train carrying the men of the "Ninth" came to a stop in the LMS Station in the Waterside. To a heroes' welcome the gunners stepped down on to the platform from where, almost five years before, they had left their native city.

Thousands jostled past the police to greet their "Derry Boys" and bid them "welcome home" and the band of the Royal Inniskilling Fusiliers struck up a musical greeting. The atmosphere was heady with emotion as the well-wishers joined in community singing to show their delight that husbands, fathers, sons, brothers and sweethearts had come home safe and well. Indeed so great was the emotion that the official welcome was almost overwhelmed, but eventually the platform party gained the attention of the crowd as the band played "For they are jolly good fellows".

Senator F J Simmons, the Mayor, told them that it was the proudest day of his life to "have the opportunity and honour of welcoming home the men of the Derry Regiment who had gone out to crush the tyrant and who came back on what they hoped was the eve of a glorious victory". (Belfast Telegraph, Oct. 20, 1944)

The Town Clerk, Mr J C Donnell, then read telegrams of welcome, including one from the Honourable the Irish Society, before Major Sir Basil McFarland stepped forward to thank the Mayor and the citizens for their warm reception. Before wishing the men good luck and a pleasant leave he expressed his own thanks for the excellent work which they had done.

Similar welcomes were accorded members of the Regiment in Limavady, Strabane, Ballymena and Belfast. Indeed their first reception had been given to them at Larne, when after a pleasant crossing on a flat, calm sea they returned to their native soil and the warmth of a large crowd on the pier singing "Come back to Erin" to greet them.

At Larne the welcoming party was headed by their Honorary Colonel, the Duke of Abercorn, Governor of Northern Ireland, and the Prime Minister of Northern Ireland, Sir Basil Brooke. The Duke welcomed them "on behalf of the people of Northern Ireland" and stated that the Regiment had covered itself with distinction. "We are all proud of you. I thank you for what you have done and wish you God speed and every good luck".

They had done "a magnificent job" said Sir Basil Brooke who continued, "as an old, but I hope not ancient, soldier, I say we can take off our hats to you boys. You

The Civic Reception in Londonderry's Guildhall for the members of 9th Regiment. October 4th 1944.
(Photo: Sir Basil McFarland's Collection)

did not ask anybody else to do the job for you; you have done it yourselves".

Over the next four weeks they were fêted at reception after reception, the biggest of which was a banquet held in honour of the Regiment in Londonderry's Guildhall on October 4th. Civic dignitaries and senior officers of all three services plus the US Forces were among the guests, who included Sir Alan Cunningham, the first Commander of Eighth Army and now GOC, Northern Ireland. Cunningham paid tribute to the work of the anti-aircraft regiments in reducing the power of the enemy air forces and made a special point of praising Irish soldiers. (He was an Irishman himself so, of course, his judgement was unbiased!)

Once again tribute was paid to the men by Sir Basil McFarland who also took the opportunity of thanking the citizens who had done so much for the Regiment's Comforts Fund particularly Mr and Mrs Lockhart, Mrs Roden and the late Mr W J Porter (he omitted to mention that he himself had contributed to the Fund). Sir Basil concluded by asking the Mayor "to convey to the Corporation and citizens thanks for that right royal entertainment and the way the Regiment had been received".

Some 1,200 people were present at a reunion dance in the Guildhall in honour of the Regiment at which the Mayoress presented a salver to Sir Basil on behalf of the parents, wives and friends of the men of the Regiment.

The Guildhall was also the venue for another special occasion, and a unique one at that, when Sir Basil McFarland was made a Freeman of the City, an honour for which he thanked the City fathers in a dignified but emotional response. As a citizen of Londonderry Sir Basil remains unique on an illustrious roll of Freemen, but few would disagree that the honour was richly deserved.

As their leave drew on the gunners renewed old acquaintances and settled for a time back into the comfort of their families. Many took the opportunity to marry and there was a flurry of weddings which included the Regimental Chaplain himself. The Rev Louis Crooks married Miss Marion Hart at Culmore Parish Church; Staff Sergeant Arthur Lynch married Miss Annie Orr at Waterside Presbyterian Church; Lieutenant Reginald Simmons married Miss Eileen McKeague at Christ Church and Gunner J J Doherty married Miss Anna Coyle at St. Eugene's Cathedral.

All too soon it was time to say goodbye again to parents, wives, children and friends. This time however there was a confidence that before long they would be reunited again and the awful spectre of war would have vanished.

CHAPTER TWO

DIVER!!!

When their leave was over the men of 9th Regiment returned to Skegness and the business of soldiering. The Regiment was now part of Anti-Aircraft Command whose Commander-in-Chief since July 1939 had been General Sir Frederick Pile, an Irishman and the son of a former Nationalist Lord Mayor of Dublin.

Training programmes began at the end of October and included arms drill and route-marching. At the end of the month came the news that the Ninth would be reformed as a mobile unit under 5th Anti-Aircraft Group and would move to Derby as part of 32nd AA Brigade.

In early November they moved into the Derby area occupying three 8-gun sites with 24 Battery at Weston-Under-Wood, 25 at Moor Green and 26 at Ticknell. Regimental HQ was established at Spondon Hall, Spondon. GL Radar Mk III which they had received but not used operationally in Italy was issued to 24 and 26 Batteries but 25 had to make do with GL Mk II. The latest Predictors were also supplied and the net effect of all of this new equipment was to make anti-aircraft fire much more effective than it had ever been before. The new radar allowed gunsites to operate independently if needed, and the number of rounds required to bring down an aircraft was reduced substantially.

New personnel were drafted in to bring the Regiment back up to its establishment, which in mid-November was changed from mobile to static. They remained in the Derby area during December, although 25 Battery moved from Moor Green to Thurlston and just before Christmas they were told that they were to go to practice camp at Weybourne in January. At the same time orders were received that the Regiment was to arrange for the medical testing of all non-tradesmen under 30 years of age. Those who were Medical Category A1 were to be transferred to the infantry. The first of those transfers took place on January 4th 1945 when twenty-two men were sent to 2nd Royal Irish Fusiliers for training and thirty-one men went to 138th LAA Infantry Training Regiment. This transfer to the infantry would take away many of the original members of the Regiment such as Jim Dinsmore, George Kernoghan and Bertie Cuthbert. Left behind were tradesmen and specialists and those over 30 years old. Pile had been forced time and again to give up trained gunners to the infantry and to reduce the strength of his Command to do so.

In 1944 the German air offensive against Britain had taken two new turnings. First there had been the "little blitz" in the spring of that year. Then on June 13th — D-Day plus seven — the first of the V1 "doodlebugs" arrived. These were codenamed "Divers" by the defenders, and a rapid deployment of defences took place to combat them. This included granting precedence to the guns and moving the gun defences to the coast. The gun zone on the south coast was named "Diver Belt" and this locked into a zone around the Thames Estuary which was called "Diver Box". However as the Allied Armies moved into the Pas de Calais area, and overran the V1 launchsites there, the flying bombs began to be launched from the Low Countries and even from

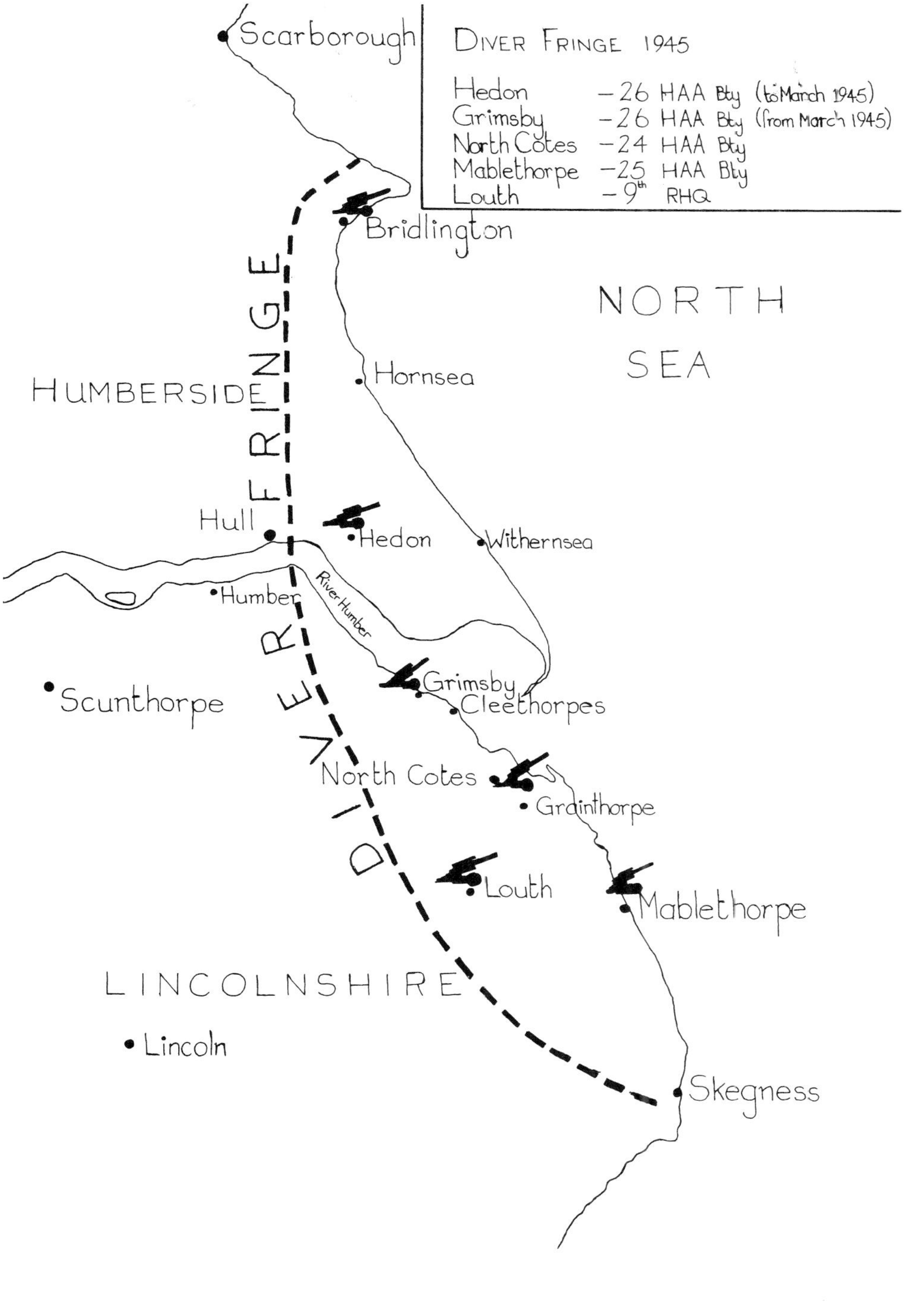
DIVER FRINGE 1945
Hedon –26 HAA Bty (to March 1945)
Grimsby –26 HAA Bty (from March 1945)
North Cotes –24 HAA Bty
Mablethorpe –25 HAA Bty
Louth –9th RHQ
Scarborough
Bridlington
NORTH SEA
Hornsea
HUMBERSIDE
DIVER FRINGE
Hull
Hedon
Withernsea
Humber
River Humber
Scunthorpe
Grimsby
Cleethorpes
North Cotes
Grainthorpe
Louth
Mablethorpe
LINCOLNSHIRE
Lincoln
Skegness

He111 bombers over the North Sea. Their path now lay over East Anglia and this caused a massive uprooting of guns from the south coast to be redeployed north from "Diver Box" towards Great Yarmouth in what was termed "Diver Strip".

Further changes in V1 flightpaths and launch areas were anticipated which could bring the deadly "doodlebugs" in north of "Diver Strip". To meet this threat an extension of the "Diver" defences, codenamed "Diver Fringe", was planned from Skegness in Lincolnshire to Whitby in Yorkshire. Shortage of manpower however meant that General Pile would not be able to put as many guns into this line as he wished and thus he shortened the line and divided it into three sections: one based on Louth in Lincolnshire; one astride the Humber and the third near Bridlington in Yorkshire.

When 9th Regiment went to Weybourne it was to prepare for deployment in "Diver Fringe", and they had to familiarise themselves with the new equipment they would be using — the Mk 2C static gun, which was favoured for the Diver role because its traverse was smoother than the mobile gun, with Remote Power Control (RPC), the new No 10 Predictor and an excellent American radar, SCR 584. This combination could literally lay a gun on to a single target. Small wonder therefore that the Mk 2C gun with such a system was referred to as "all-singing, all-dancing".

Their training programme was cut short by bad weather; that January of 1945 going into meteorological history as one of the worst on record. On January 19th 26 Battery left Weybourne for the Humber Anti-Aircraft Zone where they were attached to 152nd (Mixed) HAA Regiment (the "Mixed" indicating that the Regiment consisted of men and women, the latter being ATS personnel) as part of 65th AA Brigade. Battery HQ was at Hedon and the gunners had to prepare their own sites, as they were reinforcing the local gun defences.

Three days later 24 and 25 Batteries began deploying to sites in Lincolnshire at Marsh Chapel and Mablethorpe with a troop of 25 Battery becoming operational immediately in the "Diver" role. Two 4-gun sites were established by each battery and Regimental HQ was at Louth.

February was spent trying to improve both gun positions and living accommodation while the departure of men to the infantry continued in a steady stream.

On March 3rd at just after midnight extensive enemy air activity was noted and the guns went on full alert, but it was to prove a most frustrating night for the gunners. Overhead a force of RAF bombers returning from attacks on Cologne and the Ruhr had been infiltrated by German fighter-bombers who got into British airspace without the guns being able to fire on them for fear of hitting our own machines. The *Luftwaffe* attacked bombers as they came in to land, shooting down twenty-four, hit at airfields, road and rail transport, towns and villages and some even flew as far south as London to attack the capital. General Pile wrote later that "it was almost impossible to distinguish friend from foe in the confusion". 9th Regiment were not alone in standing powerless by their guns until they were finally "stood down" at 4.00 a m that Sunday morning at the end of the *Luftwaffe*'s Operation "Gisela".

At 7.55 p m on March 4th, 24 and 25 Batteries fired a total of 88 rounds at enemy aircraft. No hits were claimed and the enemy activity was described as being on a

A 3·7 firing at night, showing the flash from the muzzle.
(Photo: Imperial War Museum)

"reduced scale" from that of the previous night. These in fact were noted by General Pile as being the last significant efforts by manned German aircraft over Britain. Only three further small raids took place, one of which occurred on the night of St. Patrick's Day when 24 and 25 Batteries fired a total of 30 rounds. Again no hits were claimed.

On March 29th the last V1 landed in Britain, exploding in a sewage works near Hatfield in Hertfordshire. It had been one of a total of twenty-one flying-bombs launched over a twelve-hour period by German gunners from a site in the Netherlands, but the remainder had been destroyed by the ack-ack gunners on the coast or by fighters over the Channel. For AA Command the war was effectively over, and 9th Regiment had not had an opportunity to fire at a V1. Indeed in their area of operations only a very small proportion of the V1s launched against Britain had landed — two in Lincolnshire and seven in all of Yorkshire.

That they had been a deadly and effective weapon can be illustrated by the devastation produced by a single "doodlebug" landing in a housing estate in Manchester, demolishing twenty-three houses, damaging another twelve so badly that they later had to be pulled down and causing blast damage to more than a thousand other homes! In ten months V1s (and V2 rockets) had killed 6,184 civilians and 2,917 servicemen with over 50,000 civilians and servicemen injured — a third of them seriously. The Allied casualties on D-Day had been lighter!

26 Battery moved to Grimsby to take over two sites from 144 (Mixed) HAA Regiment at the end of March, and during April the Regiment's guns engaged in practice-shooting.

With the war in Europe now nearing its end states of readiness were reduced on May 2nd to ten minutes by night outside the Humber AA Zone. States of readiness were further reduced to 50% of sites manned by day on May 5th and at 8.00 p m on May 6th all but two of the Regiment's sites were declared non-operational. By May 8th all German forces in Europe had surrendered and the war on that continent was over.

Gun sites were dismantled and the guns of the Regiment began to be handed in, the first to go being those of 24 Battery which were taken by transporters to the AA Equipment Park, Arrowe Park, Birkenhead on May 19th. 25 Battery's guns followed on May 24th and the Regiment transferred to the command of 65th AA Brigade. They moved to Ludham Camp, Norfolk in early June where they were soon joined by 26 Battery who had handed in their guns on the 8th. At Ludham they took over the guns and vehicles of 157th HAA Regiment and re-organised on a mobile basis. They also had to spend a considerable portion of June fixing up and cleaning Ludham Camp which had not been occupied for some time.

On July 20th King George VI and Queen Elizabeth visited the Regiment's home city of Londonderry. Needless to say men of 9th (Londonderry) HAA Regiment[1] were included in the march past on that occasion. As they took their place proudly

[1] Sir Basil McFarland had left the army in April and had been re-elected to the Mayoralty of Londonderry by the time of the Royal visit. As Mayor and HML for the city he was responsible for some of his "Derry Boys" being brought home for this visit.

many remembered the last occasion they had marched before King George in Tripoli in 1943; little over two years had passed and yet so much had changed and the war was all but over.

During their time in Anti-Aircraft Command the Regiment had been given no real opportunity to display their prowess. Due to the advance of the Allied armies on the continent the German V1 campaign had been reduced in its intensity although, as indicated above, it had had very severe effects. The expected attacks above "Diver Strip" did not materialise and "Diver Fringe" was never fully activated. However there can be no doubt that given the opportunity the Ninth would have been equally as proficient as shooting V1s out of the sky as they had been in "downing" manned aircraft. Instead of the action to which they had become used the Regiment spent that final winter of war "standing-to" at freezing cold gunpits waiting for an enemy who rarely appeared but whose threat could not be discounted until the final surrender was signed.

CHAPTER THREE

THE BOYS COME HOME

World War Two ended completely in August 1945. The dropping of atomic bombs on Hiroshima and Nagasaki brought about the surrender of Japan, although some Japanese troops continued to fight and the British recapture of Hong-Kong after "VJ Day" was opposed by Japanese suicide boat-squads.

In spite of the large scale reduction of the forces, including of course Anti-Aircraft Command, 9th Regiment remained in being. The initial disbandment orders involved the war-raised regiments, including the "mixed" units and TA regiments. As a Supplementary Reserve unit Ninth was not earmarked for disbanding but remained to form part of the post-war AA Command.

For the original members of the Regiment however their days in khaki were drawing to a close. During the final months of their service they underwent educational programmes, assisted in local factories and in harvesting on farms, and had a surfeit of sports activities. In September they moved back to the Derby area where they had a visit from Lt Colonel Aitken, who was now commanding a medium regiment in Great Yarmouth, having been wounded in action in Italy. Mid-October saw another change of command as Lt Colonel Mayhew was discharged and handed over to Lt Colonel J O Horne. The latter was replaced soon after by Lt Colonel R J Longfield.

The "Derry Boys" were leaving too. "Demobs" had begun with the older men going first — J J Doherty who was now approaching his fortieth birthday was discharged in September — and the process continued until the spring of 1946, when all of the citizen-soldiers had returned to civilian life. (All, that is, bar a few who decided to stay on as regular soldiers.)

Jimmy Henderson moved from 24 Bty to RHQ as second-in-command, his place as Battery Commander being taken by Paddy Laird — who had been one of the regular cadre NCOs of 25 Bty in 1939. Laird's place was then taken by Arthur Hogg who had returned to the Regiment in England after being invalided home from the Salerno beaches[1]. Sir Basil McFarland had left 25 Bty in April 1945, command passing to Major Kirby and then to E H "Sonny" Babington. Unfortunately no record of 26 Bty at this time can be found.

By early 1946 the Regiment was a component of 41st AA Brigade in Anti-Aircraft Command. In the same Brigade for the first time since November 1939 was Ninth's sister regiment — 8th (Belfast) HAA Regiment which had returned from India in late 1945.

At Radcliffe-on-Trent on February 13th 1946 24 and 25 Batteries were combined although both continued to be shown as separate units in AA Command's Order of Battle.

In what might be seen as a rare flash of inspiration the War Office asked Sir Basil McFarland if he would be prepared to accept the position of Honorary Colonel of 9th (Londonderry) HAA Regiment, RA. Sir Basil accepted with enthusiasm and the appointment was published in the London Gazette on March 3rd. Lt Colonel Longfield forwarded his congratulations and a visit to the Regiment by Sir Basil was arranged. Ninth was now a regular army unit but proud of its wartime achievements and of the citizen-soldiers responsible for them. Their new Colonel exemplified the spirit of the wartime regiment and proved equally popular with the post-war gunners.

A regimental flag was proposed by Lt Colonel Longfield, who suggested that the design might include orange lilies. However, Sir Basil felt that this could be interpreted as a slight to the Catholics among the "Derry Boys" and instead proposed a design which included four green oak leaves[2], one in each corner of the flag, with their stems pointing in towards a silver "gunner" badge in the centre with the legend "9th (Londonderry) AA Regiment". The background colour of this design which was accepted was deep red.

In April 1947 9th (Londonderry) Regiment was re-numbered as 60th (Londonderry) HAA Regiment and was chosen to "parent" a new Territorial Army ack-ack regiment to be raised in Londonderry. The new TA Unit was designated 246 (Derry) (Mixed) HAA Regiment, RA (TA) and Sir Basil McFarland was appointed Colonel of that unit also.

The War Office choice of Derry for the Regiment's title was not an attempt to avoid having two units with the same territorial title — there had long been both Dragoons and Fusiliers with the territorial title Inniskilling — but was an eloquent and simple tribute to a first-class unit; the "Derry Boys".

In six long years those "Derry Boys" had progressed from being civilians in uniform with no military skills to being a superb and highly-respected regiment.

[1] Arthur Hogg's first posting after his recovery was ironically to a HAA Battery based in Londonderry. He moved with the Battery to England in the spring of 1944 and in June of that year his unit moved into "Diver Belt".

[2] The Irish name — Doire — means oak-grove and the oakleaf is symbolic of the city.

There is no exaggeration in saying that in the Middle East and Italy it was quite simply the best! Brigadier Smith of 66 AA Brigade wrote to their then Honorary Colonel, the Duke of Abercorn, on October 5th 1944 to tell him that His Grace had "*every reason to be proud of the Regimental record, of a difficult task well and faithfully accomplished and of the high reputation which the Regiment has established*". To the gunners themselves, as he said goodbye to them at Leghorn, Smith declared that they had "*covered themselves in glory*". And yet these glowing tributes seemed to rest lightly on the shoulders of the men themselves. Their modesty has been in direct proportion to their achievements; few would accept that they had behaved as heroes or even been brave. By consensus their simple claim is that they had a job to do and that they did that job as well as they could.

Notwithstanding the inherent modesty of the gunners themselves I would feel that no reader of this record could fail to agree that they were exceptionally good and that they could reasonably claim to have been "second to none".

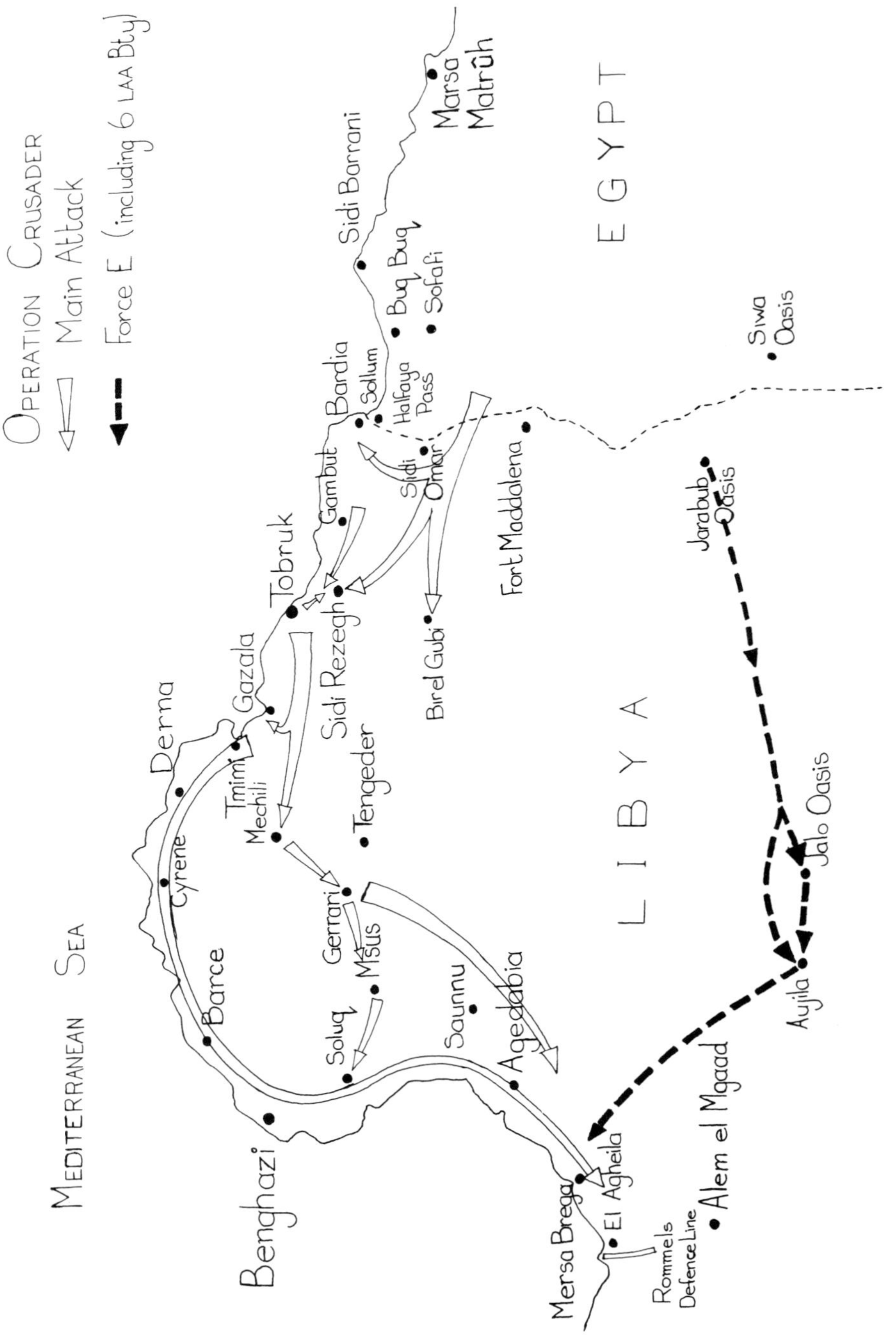
OPERATION CRUSADER
Main Attack
Force E (including 6 LAA Bty)
MEDITERRANEAN SEA
EGYPT
LIBYA
Marsa Matrûh
Sidi Barrani
Bug Bug
Sofafi
Sollum
Halfaya Pass
Bardia
Sidi Omar
Fort Maddalena
Siwa Oasis
Jarabub Oasis
Gambut
Tobruk
Sidi Rezegh
Bir el Gubi
Gazala
Tmimi
Derna
Mechili
Tengeder
Cyrene
Gerrari
Msus
Barce
Saunnu
Soluq
Agedabia
Benghazi
Jalo Oasis
Aujila
Alem el Mgaad
El Agheila
Mersa Brega
Rommels Defence Line

PART SIX

THE LIGHT BATTERIES

CHAPTER ONE

COLERAINE TO CAIRO

When the men of 9th Regiment left Northern Ireland that November day in 1939 they moved without their light battery, for the movement order did not include the Coleraine men; someone in the War Office had obviously decided that it was much simpler and cheaper to move 5 LAA Battery, which was at practice camp in Lancashire, to Hampshire than to bring a battery all the way from Coleraine.

6 LAA Bty remained in Coleraine only for a few weeks more. By the end of November they were on the move also, travelling across the channel to Scotland to join 51st LAA Regiment. There X Y and Z Troops deployed to Rosyth, Wick and Grangemouth respectively. Such was the shortage of anti-aircraft guns that they were sent to man sites occupied by telegraph poles! They did however have Lewis machine-guns with which Y Troop achieved the Battery's first "kill", shooting down a Dornier bomber at Wick.

Complete with Lewis guns they remained at various sites in Scotland until August 1940 when they moved to the LAA Practice Camp at Cark in Lancashire and from there to Minley Camp at Farnborough in Hampshire, prior to going overseas. They had two changes of Battery Commander in this period. Major William "Toby" Allison had been replaced by Major R W J Gibbon, MC in Scotland, but the latter was found to be medically unfit for overseas service and gave way to Major James Hope.

On September 8th 6 LAA Bty boarded the SS *"Dominion Monarch"* at Liverpool and sailed in convoy around the Cape to Suez, stopping at Freetown in Sierra Leone and at Cape Town. The long voyage was a very pleasant experience, "almost a holiday cruise", says Billy Oliver and it was filled with a programme of sport and training lectures. The "holiday" ended on October 19th in the Red Sea when an Italian bomber attacked the convoy, dropping two bombs near one of the ships. No damage was done although Rome Radio later claimed that the *"Dominion Monarch"* had been sunk, a claim which its passengers found most amusing.

After a week in camp near Cairo the Battery deployed for static AA defence in the Canal Zone and at Port Said. They remained in the Canal Zone until the end of July 1941 and their time there was relatively quiet. On several occasions they did open fire on enemy aircraft, but generally the attackers were too high for effective engagement by the Bofors guns. However "near misses" were noted in two raids in May and later that month there was a "flap" when the Battery was redeployed to cover the possibility of a German airborne attack as at Crete. Other than these incidents time was devoted to training and in fact the Battery had also provided a training course for Australian LAA gunners.

They had also suffered their first fatal casualty when, at the end of March, Gunner Alfred McKeown was killed in a road accident.

CHAPTER TWO

HOPE'S GREYHOUNDS

On the night of 17th/18th July 1941 6 LAA Bty claimed its first hits on enemy aircraft in the Middle East. One of Z Troop's guns had fired 28 rounds at attackers, and their initial claim of two hits was later upgraded to the destruction of an enemy plane. This was a morale-booster for the Battery and it came at a good time for they were about to depart the Canal Zone for a mobile role in the desert.

On July 21st 100 LAA Bty relieved them and the Battery moved into reserve to prepare for the desert. Twelve mobile guns were taken over from 41 LAA Bty together with gun tractors. As 41 LAA Bty had just returned from the Eritrean campaign their tractors were well worn and the Coleraine men managed to exchange them for fresher vehicles, which turned out to be the tractors they had brought out to the Middle East themselves!

The Battery's Advance Party arrived at Bagush on August 10th and was joined two days later by the main party. Under the command of 2nd LAA Regiment in 12th AA Brigade, X and Z Troops deployed to guard landing grounds near Sidi Barrani, one of which had already been protected by 25 HAA Bty. Y Troop however were ordered to move, under 2nd South African Brigade, to Siwa Oasis, deep in the desert and until their departure remained at Gerwala.

On August 15th Major Hope was appointed AADC, Barrani Landing Grounds where, as well as Z Troop at LG 02 and X Troop at LG 05, he was also responsible for 38 LAA Bty at LG 03. 6 Bty's Troops were reinforced by a number of captured enemy pieces, seven Italian Breda 20 mm guns and one German Flak 20 mm. Y Troop left Gerwala for Siwa on August 16th and four days later Battery HQ and X Troop were ordered to make the same move, while Z Troop moved from LG 02 to LG 05. Early on the morning of August 24th X Troop reached Siwa where they came under the command of 7th Indian Infantry Brigade who ordered them to prepare to move to Jarabub Oasis, a distance of 90 miles to the north-west, via Davidson's Pass. At Jarabub they came under the command of 6th Sikhs and were deployed north of the oasis.

Six SM 79s attacked Jarabub at noon on August 27th, dropping fifty bombs around the water point and in the palms. A number of bombs failed to explode although two soldiers were injured. Neither of the injured were from X Troop. By the end of the month Z Troop had also moved southwards and the entire Battery then became independent of 12th AA Brigade being assigned operationally to 7th Indian Brigade.

They engaged enemy aircraft, without effect, on six occasions during September although on two of these raids they simply fired "indicator" bursts for fighters operating from the aerodrome at Jarabub. Several Italian recce aircraft were spotted but were too high for effective engagement. Bombs fell near their B Echelon positions on September 8th causing two casualties. On September 14th there was a frantic engagement when three Messerschmitt Bf 110s and an SM 79 machine-gunned a convoy in Davidson's Pass, destroying a petrol lorry. Y Troop fired 58 rounds and claimed one hit while Z Troop fired 5 rounds with no hits.

During October the area received considerable attention from recce aircraft, thirteen overflights being noted but none were in range of the Bofors and thus the Battery's guns remained silent. The guns of Y Troop under Lt Butler moved out to accompany South African armoured cars on a five day desert patrol on October 21st and their place was taken by dummy equipment. By now the Battery was under the command of Major J G Siderfin, who had replaced Major Hope on the latter's appointment to an AA staff post. Hope left a permanent legacy behind, for 6 LAA Bty, on account of its desert wanderings under his command, had achieved the soubriquet of "Hope's Greyhounds".

Only an occasional high flying recce plane disturbed the peace at Jarabub in early November but then on November 15th, just a few days before the launch of Operation "Crusader", the oasis and its aerodrome came under heavy air attack. The attack, by SM 79s and Ju 88s, started at 9.05 a m; while the bombers operated at high level, Bf 110s came in low, bombing and strafing, through the smoke and dust raised by the bombing. A small fuel dump and a Blenheim bomber were destroyed but one of the attackers was hit by Sergeant Sam Pinkerton's gun and subsequently crashed. X Troop was not involved in this action, having left Jarabub on a Brigade desert formation exercise just two hours before the raiders struck.

Y and Z Troops, accompanied by a small BHQ, left Jarabub on November 18th, the day before "Crusader" began, to protect a mobile force designated Oasis Force which was to attack Italian forces on the southern flank of the Axis held area.

Initially Oasis Force made good progress but they then began to encounter small wadis and limestone depressions which impeded them and caused a number of problems which included one of Z Troop's tractors breaking two crown wheels. Then at noon on November 21st the column was attacked, as it moved through a wadi, by Junkers and Fiat bombers and Bf 110 fighters. The 110s machine-gunned one of the Battery's 3-tonners which had taken the stricken gun-tractor in tow. Both vehicles were destroyed and Gunner Paddy Toner, a native of Coleraine, was killed trying to get clear of his lorry. The guns were rapidly brought into action and a hit was claimed on one bomber before the engagement ended. However their problems were not yet over as the soft sand slowed them considerably and the tractors had to be towed by the 3-tonners.

Next day the column split for the attacks on Axis positions at Jikara and Jalo. Lt Martin Jackson, with the guns of Sergeants Bertie Currie and Sam Pinkerton, was

The remains of the Fiat BR 20 bomber shot down by Sgt Jim Murray's gun on November 24th lie near Jalo Oasis.

(Photo: 6 LAA Battery OCA)

Soft going. The difficulties experienced in movement in the desert are well illustrated by this gun and tractor of 6 LAA Bty during Operation Crusader.

(Photo: 6 LAA Battery OCA)

attached to Colonel Ghroebblar's column for the assault on Jikara. So bad was the terrain however that the guns were left to the rear of the column, which again was bombed and strafed but with no casualties to the gunners. Another group, under Lt Brian Clark, accompanied the Brigade party to Augala which they entered that evening, having been bombed once enroute.

The force maintenance column, supported by Y Troop, remained at Bu Etla. For a period of four hours from 11.30 a m this column suffered almost continuous bombing and strafing attacks. Y Troop claimed hits on two aircraft as well as shooting down a Fiat bomber. In fact the guns appeared to be the main target of the attackers and the crews were totally exhausted by the end of the day, as was Gnr Ron Hutchinson who was commended for his excellent work in bringing up ammunition in his 15 cwt truck throughout the action.

At dawn next morning the maintenance column moved off but, early on the journey, Y Troop's guns bogged down and had still not been extricated by nightfall. However both Bertie Currie and Sam Pinkerton's detachments had caught up with Ghroebblar's column near Jikara where enemy guns covering open ground had forced the column to stop. As the field guns were also bogged down some distance away Currie brought his Bofors into action against the Italian guns. Lt Jackson went forward to act as observer and all of this time the gun was in full view of two 47 mm and two 20 mm guns. Currie's second burst of fire scored a direct hit on one of the 47 mm guns and the other guncrews promptly surrendered, allowing the infantry with armoured car support to advance and take Jikara. For his courage in this action Martin Jackson was awarded the Military Cross.

Z Troop's remaining guns accompanied the Brigade force through the night from Augala to Jalo where at 8.30 a m on November 24th they were attacked by twelve CR 42 fighters and two BR 20 bombers. In the ensuing engagement both Sgt Robert Burns and Sgt Jim Murray's guns shot down an attacker apiece — a CR 42 and a BR 20 respectively. At midday they were joined by Jackson and his two guns which had left Jikara at dawn. Jackson's guns immediately took cover in scrub but were bombed by enemy planes and later shelled by artillery from the fort at Jalo. The fort was however captured by an infantry attack that same evening.

Y Troop's guns and several vehicles which they salvaged along the way arrived at Jalo from Bu Etla via Augala the following morning after an eventful journey. A 15-cwt had been machine-gunned but, fortunately, no-one was injured.

The guns were dug in for much of December which opened with a bombing and strafing attack from which no casualties resulted. Guns from Y and Z Troops accompanied offensive patrols in the middle of the month and then on December 20th BHQ and Y Troop left Jalo with the main body of Force E in a north-westerly direction towards Mersa Brega. Lt Jackson set out with two guns of Z Troop and a number of armoured cars to clear the country ahead of Force E. After some soft going the way got easier and fairly good progress was made, but next day soft sand and camel scrub slowed both ack-ack and field guns. An improvement in terrain enabled the guns to rejoin the main body that evening and the following morning they occupied positions around the fort at Geof-el-Matar which had just been captured. Later that same morning (December 22nd) two Bf 110s strafed the vehicles setting one lorry on fire.

Patrols were sent out from Geof-el-Matar accompanied by guns of the Battery for AA defence and Christmas was celebrated with the only concession to festivity being an extra rum ration that night. Additional defensive measures were instituted on December 26th following news of enemy forces moving towards the area.

BHQ, Y and Z Troops left Force E at Geof-el-Matar on January 17th to join 22nd Guards Brigade at Samhaudia, south of Agedabia, leaving X Troop detached at Geof-el-Matar. From Samhaudia, on January 20th, Lt Butler with three guns left to join Sainthill Column while the remaining five guns stayed in the B Echelon area of 22nd Bde. Next day these latter guns engaged six CR 42s which attacked a convoy passing the B Echelon position. The fighter pilots had obviously not noticed the Bofors, for two of the Italian aircraft were shot down and another two damaged. In retaliation the Italians strafed the Bofors but failed to cause any casualties or damage, other than puncturing a lorry's radiator. An hour later, at noon, enemy columns were spotted and at 2.00 p m an order was issued to retire northwards for twenty miles.

A further withdrawal was ordered a day later, and they began to be concerned for the safety of the men with Sainthill Column. As they retired they were aware of a tank battle raging to the north of the Guards Brigade and they skirted this action while passing through Wadi Saunnu where the Battery was cut off from the column. North-east of Saunnu 6 LAA Bty halted for the night but there was little opportunity for rest as tanks and other vehicles were passing close by on either side.

They rejoined the Guards Brigade at midday the following day. Shortly afterwards Butler appeared with his three guns, departing almost immediately with an additional weapon.

With Rommel now advancing vigorously the withdrawal continued and the B Echelon pulled back towards Mechili where part of X Troop joined up, having become detached from Force E during the retreat. At Mechili six of the Bofors were placed under the command of Lt-Colonel Sampson of the South African Field Artillery to act as anti-tank guns while efforts were being made to locate the main body of 22nd Bde. Eventually orders were received to rejoin the Guards at Charruba. The six guns moved up towards Charruba on January 29th with B Echelon of 22nd Bde and two days later contact was finally made with the Brigade astride the track at Ras Nefiet. Y Troop was found to be intact, having had an adventurous time and another X Troop detachment had rejoined also.

Sgt Billy Oliver's detachment was however missing and it was later confirmed that they had been captured by the Italians near Agedabia. Taken with Sgt Oliver were L/Bdr Jim Hamilton, Gnrs Jim Doherty, Jack Backshall, Teddy Blackburn, Robert Cairns and John Boyd. L/Bdr Willie Dysart managed to escape and later rejoined the Battery.

It was also believed that Sgt David Pringle and his detachment as well as Troop HQ had been captured or killed and fear for their safety grew as BHQ moved first to Mechili and then, on February 3rd and 4th, back to Gazala. There, however, the missing men returned in the company of a patrol of the Long Range Desert Group under Irish rugby international, Cecil Beamish. Davie Pringle still believes that a full ration truck was the real reason for the LRDG men staying with the gunners, although Beamish would no doubt have been pleased to stay with the Coleraine men, having gone to school in the town.

6 LAA Bty left the Guards Brigade on February 7th to join the Polish Brigade under Major General Stanislaus Kopanski in the Carmuset-er-Regem area south of Gazala. At 8.00 a m on February 8th a number of Junkers Ju 87 dive-bombers passed overhead too high for effective engagement. The Poles however expended copious quantities of small arms fire on these and other high-flying aircraft! 6 LAA Bty's gunners remember this fire as being almost totally ineffective and even occasionally a danger to the Poles themselves, and to the gunners, whenever machine-guns were fired at low angles of elevation! In deference to Polish wishes the Battery agreed to extend their engagement range even though they knew that this would be a waste of ammunition.

The Polish field gun positions were heavily dive-bombed next morning bringing the Bofors into action against the attacking Stukas, two of which were shot down while a "probable" was also claimed. Sgt Currie's gun was credited with one "kill" and the other was shared. However Sgt Bertie Dunlop was wounded in the right lung by a bomb splinter and later died from his injuries.

Twenty-four hours later two gunners — Harry McWilliams and William Campbell — were wounded in another attack by twenty Ju 87s. Sgt Currie's crew claimed a hit but no planes were shot down.

No further enemy air activity occurred until February 15th when three Bf 109s were fired at without effect at 8.30 a m. Heavy enemy shelling fell on the area later that morning forcing South African armoured cars to withdraw. Lt Jackson found himself ordered on patrol with three Bofors guns, a like number of Polish field guns and six Bren gun carriers. It was a confusing episode — the Poles seemed to have no real idea of why they had sent out the patrol and Jackson found himself in receipt of conflicting orders from different Polish officers.

Sgt Currie had a narrow escape during this patrol when his 3-tonner was struck by a dud 105 mm shell! The patrol witnessed the destruction of an RAF Kittyhawk fighter by two Bf 109s. The pilot, Squadron Leader EM Mason, was killed and his remains were removed from the wreckage of his plane for burial by the gunners.

A week later Jackson and Lt Clarke were sent forward on two occasions to cover 234 Bty of 68th Medium Regiment which was shelling enemy positions at Tmimi and at a point five miles north of Carmuset-er-Regem.

Then on February 26th three Bf 109s fell to their guns. The first was destroyed by Sgt McCaw's gun and crashed some ten miles south-west of the Battery's position. That evening two Bf 109s which flew across the area at 20 feet were engaged by all the guns. One crashed close to Sgt Walls' gun and the other came down to the north-west. So many claims were made for this last machine that the Battery War Diary wryly notes — "hits claimed by everyone except BQMS Cuthbert"!

CHAPTER THREE

BACK TO ALAMEIN

6 LAA Bty remained at Carmuset-er-Regem until March 24th during which time they were engaged against enemy aircraft on six occasions. These attacks were a mixture of strafing and dive-bombing sorties, and hits were claimed on eight aircraft. Of these, three were destroyed — a Ju87 "Stuka" and two G50s. They were relieved by 6 (South African) LAA Bty and ordered back to relieve 5 (SA) LAA Bty of the defence of the desert railhead. This handover was gradual as the two units also exchanged guns — 6 LAA Bty's Bofors for 5 (SA) LAA Bty's Italian Breda guns.

In early April six Bredas were moved from Capuzzo to protect the desert railway between stations K103 and K125. The guns were mounted on railway wagons and scored their first victory on April 2nd when Sgt Murray's gun shot down a Bf109, although Murray himself was wounded twice by ricocheting bullets. Throughout the month a number of attacks were made on trains and stations by German and Italian aircraft. The Battery opened fire on a number of occasions but no further hits were claimed.

In May there were some heavy raids on the railhead and on trains. C Troop fired 1,000 rounds on May 7th and 1,087 rounds next day. There was some damage to trucks on the first occasion, but little on the second. However a bombing attack by one aircraft on May 26th saw 23 wagons burned out when a petrol train was hit. The following day Sgt John Hunter's gun fired 24 rounds at a low-flying attacker which subsequently crashed.

Just after midnight on May 28th the Battery achieved a unique distinction for an AA unit when it captured an enemy aircraft intact! The events leading to this capture began when a convoy of the Royal Indian Army Service Corps heard an explosion along the railway line. An officer went to investigate and two guns under Sgt Hunter and Sgt Alan Elliott went up on their portees. They found a light plane — a Fieseler Fi156 Storch — beside the line with its engine ticking over. There was no sign of the crew, but the RIASC officer later said that he believed there had been two aircraft and that the infiltrators had all escaped in one. Hunter thought that he had heard an aircraft take off as he approached, and there is little doubt that the swift arrival of the guns thwarted the daring sabotage mission which had been in progress. The damage which had been done was repaired in a matter of hours but the saboteurs, if left alone, would have been able to blow up an important trainload of tanks! The Storch was later handed over to the RAF.

In June Major D J (Jack) Christie, an original officer of the Battery, took command as they continued to protect the Western Desert Railway, both alongside and on its tracks. Towards the end of the month BHQ, A and C Troops (the Troops had been redesignated A, B and C some months before) moved back to Mersa Matruh with B Troop remaining at Mischiefa under Captain Butler.

A hit was claimed on an attacker at Matruh by C Troop on June 24th. Next day BHQ, A and C Troops moved to Fuka where, during a heavy raid, A Troop claimed

Booty!! The Fieseler Storch which was captured by Sergeants Hunter and Elliott provides a backdrop for a photograph of the captors before their prize is handed over to the RAF.

(Photo: 6 LAA Battery OCA)

General Montgomery pictured with Sergeant Bob Balmer's gun crew at Benghazi on November 29th 1943.

(Photo: 6 LAA Battery OCA)

at least one hit. Night-fighters shot down two aircraft and there was no damage on the ground. B Troop had moved to Simila in the meantime and then to El Imayid. Elements of the Battery moved to El Hamman to defend its railway station while B Troop moved to Burg-el-Arab. With Rommel bearing down fast on the El Alamein position all 'soft' vehicles, stores and spare personnel were ordered back to the RA Base Depot in case the Battery should be forced to make a rapid retreat.

During the first Battle of El Alamein 6 LAA Bty moved to Amiriya, with C Troop defending the railway station and A Troop the causeway, before deploying to Kilo 25 on the Alexandria road to provide air defence from Mena House to Kilo 10. BHQ was situated near the Great Pyramids.

Mobile training filled their time for the remainder of July with A Troop moving to defend Heliopolis aerodrome on July 30th. A raid by about twenty planes on the aerodrome at 2.30 a m on August 1st saw A Troop fire 1,100 rounds and destroy two enemy aircraft. Four RAF men died in the raid. Later that day BHQ, B and C Troops moved to Heliopolis. B Troop went to Abbasia three days later and on August 21st C Troop moved to Helwan aerodrome. The Battery was now under the command of 37th LAA Regiment.

CHAPTER FOUR

THE BIG PUSH

During September 1942 6 LAA Bty was dispersed, with BHQ and A Troop at Heliopolis, B Troop around Abassia and C Troop at Helwan. Training continued in preparation for the "show" that was coming up, and then on September 25th Heliopolis again came under attack. In the course of this raid which began at 8.18 p m the guns engaged the attackers with barrage fire, putting up a total of 1,965 rounds. Twelve bombs were dropped but no damage was caused to the aerodrome or its aircraft. However the Battery suffered two casualties with both Major Christie and the Battery Captain, Robin Wray, being wounded by shrapnel and admitted to hospital. Lt William Dickson assumed temporary command until Major Christie's return. In early October they received the thanks of the RAF Station Commander at Heliopolis for their excellent work during the two raids on the base.

An additional Troop was created as the result of a reorganisation of the Battery into a four troop unit[1] — each with three guns — on October 12th. The newly-formed D Troop, commanded by Lt Bob Pelham left the Battery to join Tactical HQ (Tac HQ)

[1] The reorganisation into 3 gun troops came about as the result of problems with the standard deployment of the 4 gun troop. This had involved a gun being placed in the middle of a triangle each apex of which was formed by another gun. In practice the centre gun had been unable to operate effectively at all times due to the dust and smoke from its companions.

Eighth Army for the protection of the army commander — General Montgomery. A day later, under command of 2nd LAA Regiment, they moved to positions at Bahig for railway defence and were still there when the Battle of El Alamein opened on October 23rd. 127 LAA Bty relieved BHQ, A and C Troops at Bahig on October 26th, the same day that Major Christie returned to his Battery.

During November the entire Battery was involved in the advance westwards, moving up to El Alamein and El Daba and on to Capuzzo and Tobruk. As they passed through the Halfaya or "Hellfire" Pass on the way to Tobruk the convoy in which they were travelling came under attack from enemy aircraft. The driver of a 5-tonner in front of the Battery's vehicles abandoned his lorry so quickly that he neglected to put on his handbrake and the lorry began to roll back towards the following vehicles. As these included a number of petrol tankers the result could have been catastrophic had it not been for Gunner George Johnston, who slewed his gun tractor across the road, leapt out, jumped into the careering 5 tonner, and prevented it from crashing into the petrol tankers. For his quick-witted and gallant action George Johnston was awarded the George Medal while Sgt Frank Walls received a Mention in Despatches for his part in the incident.[2]

Martuba was reached on November 22nd but not without incident, for on that morning Major Christie's car had been hit by splinters from a bomb which burst nearby. Gnr Hugh Lynn was wounded in the head and shoulders but neither the Battery Commander nor his driver, "Butch" Atherton, were injured although the car was a "write-off". That evening the Battery was deployed for the defence of units passing through the Barce Pass and next morning the first elements reached Benghazi where harbour defence positions were taken up on November 24th.

A raid by one plane on November 25th caused neither damage nor casualties and another single aircraft was engaged with 62 rounds four days later. At 4.00 p m that day (November 29th) "Monty" paid an unofficial visit to the Battery, stopping to be photographed with Sgt Bob Balmer and his crew.

D Troop rejoined the Battery at Benghazi in mid-December — a month which saw a number of raids on the harbour. 913 rounds were fired during the first raid on December 4th, and then on December 7th A Troop claimed hits on a Ju88, one of ten bombers to attack the harbour.

Between December 13th and 15th a total of 3,194 rounds were fired in three raids, in the last of which a Ju88 was shot down by Sgt Walls' gun. Single bombers attacked on December 17th and 18th, and on 19th at least six machines came over in a determined attack during which the searchlights illuminated several targets and a number of hits were claimed. Two more raids occurred before Christmas, during one of which four aircraft were engaged with 1,415 rounds.

Christmas itself was in marked contrast to that of a year before, with Divine Service in the morning, a good meal, and a football match between the gunners and the officers in the afternoon; the gunners won by two goals to one.

Fear of a possible seaborne attack by the Germans resulted in ground defence

[2] For details of this incident I am indebted to Sam Anderson of 6 LAA Bty OCA who witnessed it.

schemes being worked out during January 1943, but the Battery continued to operate in its normal role and four raids were engaged with a total of 1,148 rounds being fired. A He111 fell to the nightfighters during one raid.

BHQ with B and C Troops were ordered to Bercis on February 3rd for the defence of a landing-ground occupied by 89 Squadron RAF and Major Christie was appointed AADC, Bercis taking over temporary command of 2nd LAA Regiment at the same time. Captain Gibson assumed command of 6 LAA Bty.

The aircraft presence at Bercis grew to some fifty machines — Beaufighters, Hurricanes and Spitfires — of 89 and 33 Squadrons, RAF and 7 Squadron, South African AF before the Battery was relieved by 124 HAA Bty on February 20th. The month had been quiet with no enemy air activity and this pattern was repeated in March, during which they moved up towards Tripoli for the defence of the Misurata landing-ground under 2nd AA Brigade — the same Brigade which "owned" their parent regiment, the "Ninth".

April was also free of Axis attacks so that 6 Light's time at Misurata was in marked contrast to that of their fellows in 24, 25 and 26 Batteries at Tripoli. They reverted to being a three Troop unit on April 9th, a day before twenty-four men were posted in from 9th Regiment. These were original members of the Regiment who became surplus to establishment and whose transfer to 6 LAA Bty had been "engineered" so that they could remain in a unit in which they might feel at home.

There was one strafing attack on May 17th when a Ju88 flew low across the area but no shots were fired as there had been no early warning. The Battery eventually moved to Tobruk, where they arrived on June 12th. From Tobruk guns were deployed to the railhead, to Gazala, Martuba and Appalonia. With no enemy air activity the emphasis was on training, although this was curtailed by the Battery's movements from location to location. Further moves saw them at Derna and El Adem in July while in August some guns were moved to Bu Amud and Gambut with C Troop up at Derna. Under the command of 96th HAA Regiment they concentrated in the Tobruk area in mid August and were deployed to protect US aircraft.

The order to leave North Africa finally came in November and they travelled in a convoy from Alexandria through the Straits of Gibraltar to Greenock, where they arrived on December 9th. A month of disembarkation leave meant that Christmas 1943 was spent at home, where they were warmly received by their families, friends and community.

CHAPTER FIVE

THE LIBERATORS

In January 1944 the forces which were to invade France later that year were being assembled and trained in Great Britain, and it was to the British Liberation Army that 6 LAA Bty was assigned when its men returned from leave. The Battery was regimented with 113 and 149 LAA Btys in 27th LAA Regiment and was based at Ipswich to carry out its training programme for Operation "Overlord".

The Battery went on exercise in the Thetford Battle Area in April before moving to Clacton in May for the waterproofing of its vehicles and equipment. After this it moved into the marshalling area and eventually sailed for France from Tilbury on the SS "*John Souter*". Disembarkation took place at Le Hamel near Arromanches on July 4th, and the Battery moved into the concentration area for 30 Corps in the British 2nd Army.

Their first deployment was along the road from St. Martin-les-Entrées to St. Leger in their normal air defence role, but they had been ordered not to fire at any aircraft unless it committed a "hostile act". Apparently the Army Commander had nearly been shot down by LAA guns of the RAF Regiment and had subsequently issued the order restricting the actions of all his AA resources. Thus when the Corps Commander, Royal Artillery (CCRA) visited the Battery, Major Jack Christie expressed the view that "we are not earning our pay". Brigadier Rawlins, the CCRA, soon remedied that problem for two of the Battery's Troops (A and C) were ordered to convert to being armoured within three days. Christie asked for and got an extra day and on July 16th A and C Troops went into action with four Centaur tanks and drivers from the Royal Armoured Corps. The tanks were "clapped out" but their guns were still effective and, in the circumstances, that was what counted.

The Centaurs were armed with 95 mm howitzers — a weapon developed from the barrel liner of a 3·7 AA gun and the breech of a 25 pounder — designed for close support work. A and C Troops were to operate with the Divisional Artillery in a counter-mortar role until July 30th, and it proved to be a very busy time indeed, with their Centaurs being described at one time as the "most forward artillery in the British army". Inevitably they themselves came under counter-battery fire from the German artillery but fortunately there were no casualties or damage. Because of the limited elevation of the guns the Centaurs had to be nosed up onto banks to get the elevation required for shooting on many occasions. At one point they even had to resort to using railway sleepers for this purpose!

With the Battle for Caen over, 30 Corps moved towards Falaise to assist in the closing of the Falaise Gap. It took until August 20th finally to close the Gap, at a terrible cost in Allied casualties and with a controversy over strategy that persists to this day. During the Battle 6 LAA Bty not only carried out their primary function of air defence but also served as field gunners, using their Bofors guns to good effect against enemy vehicles and ground forces. On August 21st the Battery were at Argentan, from where they moved to L'Aigle and then to Pacy-sur-Eure, which was

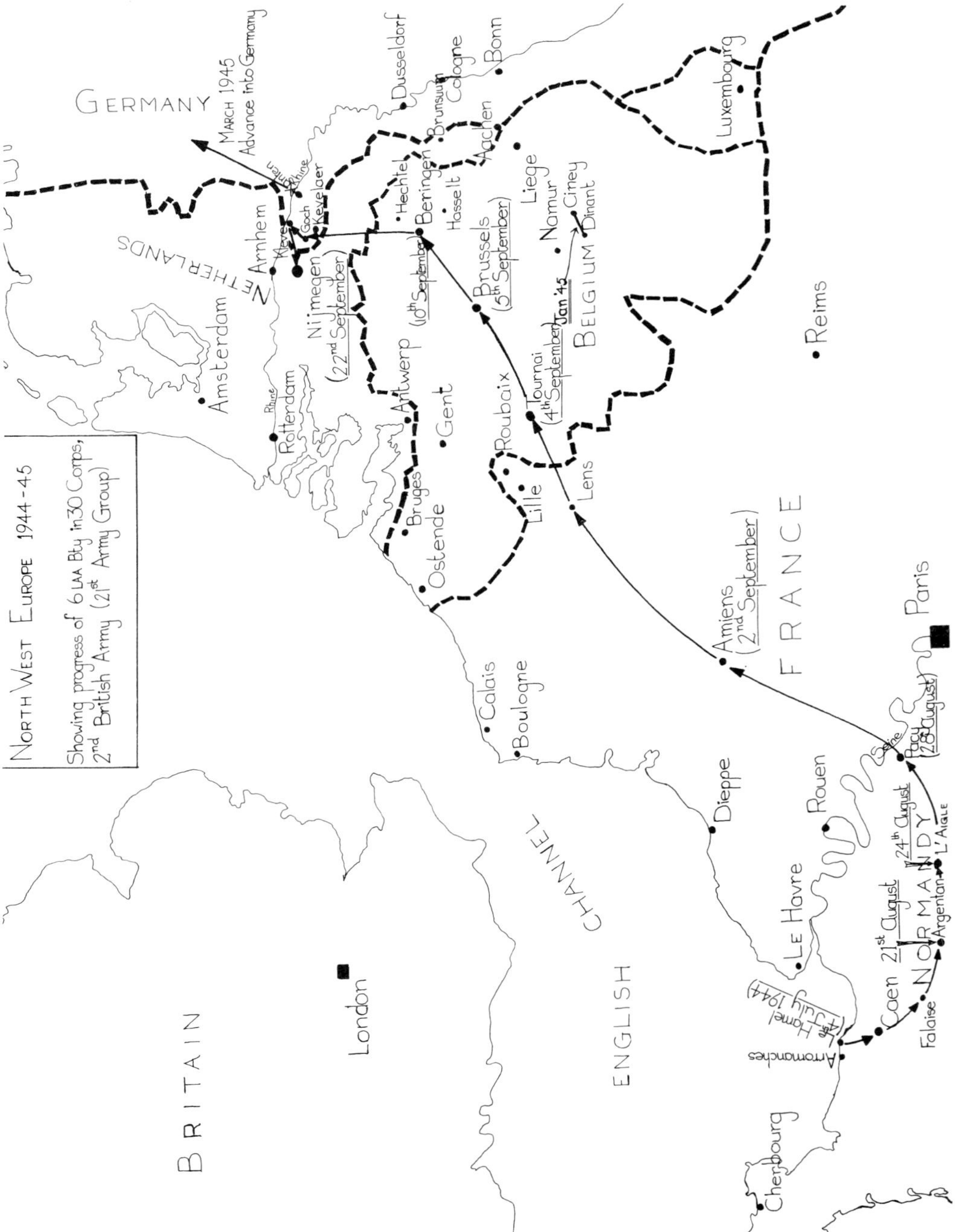
North West Europe 1944-45
Showing progress of 6 LAA Bty in 30 Corps, 2nd British Army (21st Army Group)
Germany
March 1945 Advance into Germany
Netherlands
Belgium
France
Britain
English Channel
Normandy
Luxembourg
Dusseldorf
Cologne
Bonn
Aachen
Brunssum
Liege
Namur
Ciney
Dinant
Jan '45
Hasselt
Hechtel
Beringen (10th September)
Brussels (5th September)
Tournai (4th September)
Roubaix
Lille
Lens
Amiens (2nd September)
Reims
Paris
Pacy (28th August)
Seine
Rouen
Dieppe
Le Havre
Calais
Boulogne
Ostende
Bruges
Gent
Antwerp
Nijmegen (22nd September)
Arnhem
Kleve
Goch
Kevelaer
Rhine
Amsterdam
Rotterdam
London
Cherbourg
Arromanches
Hamel (17th July 1944)
Caen
Falaise
Argentan
L'Aigle
21st August
24th August

reached on August 28th.

Paris had already been liberated and for 30 Corps and Second Army the next objective was the city of Amiens and the bridges over the Somme. Amiens fell on August 31st and 6 LAA Bty remained in AA defence of the area and the river crossing until relieved by 121st LAA Regiment on September 2nd. They continued to move north-eastwards via Lens to Tournai where they were again deployed in the AA role on September 4th. Next day they moved to defend a landing ground of the Allied Tactical Air Force near Brussels which had been liberated by the Guards Armoured Division led by Sir Allan Adair on September 3rd.

From Brussels the advance took them to defend another bridge at Beringen where they came under shelling but without suffering any casualties. Another role-change followed — to anti-tank — as the onward march of the Allied forces continued towards the Netherlands. On September 10th C Troop, equipped with seven armoured cars, was placed under the command of a Dutch Brigade.

One of the most daring operations of the war was about to be launched. Operation Market Garden saw the Allied 1st Airborne Army drop at Eindhoven, Nijmegen and Arnhem, while armoured forces raced along a narrow corridor to reinforce them and exploit their gains in order to force a rapid break-through of the German lines. The Coleraine gunners were now part of the ground forces in the race for Arnhem. The objectives for the airborne troops — British 1st Airborne and US 82nd and 101st Airborne Divisions — were the bridges at Nijmegen, Grave, Veghel, Eindhoven and Arnhem. On Sunday, September 17th the airborne soldiers went in by parachute and glider while 30 Corps spearheaded Second Army's dash along "Hell's Highway" to link up with their airborne comrades and to force, in Montgomery's words, a "thrust to the heart of Germany".

For the ground forces the race towards Arnhem was an unparalleled experience. Jack Christie maintains that his Battery got through only because the men were so enthusiastic while Bob Balmer recalls that the Battery was actually cut off by the Germans between Veghel and Grave[1]. On the road to Nijmegen Christie saw a truck in front of his jeep blown up by a direct hit from a shell. He was certain that his own Despatch Rider had been killed and was amazed, and relieved, to see the Don R ride out of the cloud of debris. Miraculously the Battery had no casualties and reached Nijmegen safely.

On their way to Nijmegen they had provided AA defence at Veghel and at Neer Bosch before taking up a similar role at Nijmegen, where the bridges had been secured on September 20th. However Arnhem was not relieved due to a combination of factors on which much has already been written. It is sufficient to say that the men at Arnhem and those trying to link up with them did all that could be expected of them.

At Nijmegen the Battery shot down a German fighter which had been lured into range of the guns by an AOP pilot — an Irishman called Steve Boycott — who set himself up as a decoy. As the German tried to shoot down Boycott he himself was

[1] The Germans did succeed in breaking through Lt General Horrocks' 30 Corps in two places and 6 LAA Bty was between these points. 30 Corps however repelled both attacks thus allowing the advance to continue.

blown out of the sky by the guns of Sergeants Tommy Henry and Percy Montgomery. Boycott was a good friend of Jack Christie and the decoy idea had been discussed by them beforehand; it was "like duck-shooting" the Battery Commander recalls.

During October the Battery was deployed in defence of other formations including field and HAA batteries and on November 11th moved to Brunssum, with C Troop deploying to Grotenrath on December 5th.

Eleven days later two *Panzer* Armies smashed through the American front line in a dash for Antwerp. Supported by the *Luftwaffe*'s last major operation they forced the Americans to withdraw and on December 21st 6 LAA Bty was moved to Hasselt to prepare for a joint AA/anti-tank role. As the "Battle of the Bulge" continued the Battery moved to Namur under the command of 6th Airborne Division and was deployed along a line from Ciney to Dinant for much of that bitterly cold January.

Early February saw them move back to Eindhoven, from where they took part in the advance into the Reichswald — Operation "Veritable". After a period of AA defence work part of the Battery converted to field artillery firing rockets in the Xanten area. Jack Christie was Forward Observation Officer and describes the sound of the rockets passing overhead as being "like ten express trains". The Germans at whom the rocket fire was aimed gave up and came out "dazed, didn't know what had hit them".

On March 21st C Troop deployed under the command of 113 LAA Bty for Operation "Plunder", the Rhine crossing which took place two days later. Although heavily shelled C Troop suffered no casualties and in fact contributed to 113 Bty's claim of two enemy aircraft "probably destroyed".

The Battery concentrated at Wesel at the end of March and in early April converted to infantry to perform guard duties on various installations and Displaced Persons Camps until the end of the war.

In June, after handing in their guns at Hamburg they took over guarding a former concentration camp which was now being used to house German prisoners. At the camp at Sand Bostel they had a Soviet liaison officer, Captain Popkoff, who had been a prisoner in the camp himself and had been sent back by the Red Army to arrange the erection of a memorial to the Russians who had died in the camp.

When he went to Sand Bostel with an advance party in May Jack Christie had the harrowing experience of discovering another, unattended, former concentration camp. Opening the door of one of the huts he was met by the appalling stench and sight of a pile of putrefying corpses. When he reported this find to the medical experts they ordered the camp to be burned to the ground immediately, an operation which was carried out by flame-thrower tanks.

Their guard duties at the camp ended in October when the Battery moved to Ottendorf. By now of course "demobs" were taking place, and in 1946 6 LAA Bty returned to the United Kingdom where it was finally disbanded.

AUTHOR'S NOTE

It would appear that no separate War Diary was maintained for 6 LAA Battery from January 1944 onwards. The Battery is included in the War Diary of 27th LAA Regiment with which it was regimented

On parade in England prior to the invasion of Europe. A self-propelled Bofors gun drives past. Note the white star of the Liberation Armies on the door and on the bonnet; inside the star is a shamrock. Major D J Christie, 6 LAA Bty's Commander, is the officer to the right of the picture.
(Photo: 6 LAA Battery OCA)

At readiness on the Dutch/German border, December 1944. The gunners remain on the alert as the fighters "mix it" above, decorating the sky with their vapour trails.
(Photo: Imperial War Museum)

and the Regimental Diary contains less detail than would a Battery Diary. I am however grateful to Lt Colonel Jack Christie for his invaluable help in completing this section of the history of 6 LAA Battery; no-one else remembered that the Battery was in 27th LAA Regiment and without this information my task would have been very difficult indeed and the end-product most unsatisfactory.

CHAPTER SIX

THE RATS OF TOBRUK

On formation 3rd (Ulster) Anti-Aircraft Brigade did not include a Light AA regiment, although its Searchlight Regiment later converted to this role. However each of the two AA regiments had a Light battery in addition to their three heavy batteries. These were 5 LAA Bty in 8th (Belfast) Regiment and 6 LAA Bty in 9th (Londonderry) Regiment. 5 LAA Bty was recruited from around the north Down and Ards Peninsula area and was commanded by Major Karl Webb. The first gunners were sworn in on April 25th 1939 and recruiting was brisk.

On August 24th 1939 the Battery was embodied and moved into war positions in the Belfast area, with its Battery HQ at Belfast Harbour Airport. The strength of the Battery was 146 officers and men and its equipment was eight Bofors 40 mm guns — four of which had been taken over from 6 LAA Bty.

In mid-October the guns were withdrawn for shipment and on October 18th the positions were handed over to 175 LAA Bty of 102nd AA Regiment — another local Supplementary Reserve regiment which had been formed as recently as August. Three days later the Battery sailed from Larne to Stranraer and travelled down to a practice camp at Cark-in-Cartmel in Lancashire where they remained until November 5th, when they moved south to Blackdown Camp to join 9th Regiment as their Light Battery. With the "Derry Boys" they went to Egypt and the joys of Sidi Bishr Camp.

However they had left two of their junior officers behind. The Battery Captain had decided that these two men were not fit enough to be officers and both were sent back to Belfast. One later transferred to the Royal Air Force and earned a Distinguished Flying Cross, while the other went to the Royal Ulster Rifles from which he transferred to a Commando in the Middle East and from there became one of the founder members of the Special Air Service Regiment. Ted Griffiths and Blair Mayne were the two individuals concerned and their war records put the lie to that early assessment of their abilities! It is believed that the man who tried to damn them ended the war as he began it — in the rank of captain.

Although the Battery chose positions for their guns in the Alexandria area they had to wait until May 15th 1940 before the weapons were finally delivered. Indeed they had to borrow a Bofors gun from the Egyptian army for training purposes in April.

When the guns did arrive there were no gun stores with them but none the less they were quickly moved into position with two troops — Y and Z, with eight guns

— in Alexandria itself and with one troop — X, with four guns — at Aboukir. The guns in "Alex" were sited for the defence of the harbour area, including the floating dock, and for the oil storage tanks at Gabbaria. Two guns were actually placed on the outer breakwater with the task of anti-minelaying defence. X Troop's guns at Aboukir were sited for the defence of the aerodrome there. 5 LAA Bty took part in none of the engagements during the raids of June 1940 as the attacking aircraft were generally at heights well above the effective ceiling of the Bofors guns.

In July they finally moved out of Sidi Bishr to Masarita near El Silsila, and their first action came on August 15th when, at last, some of the attacking aircraft flew low enough for the Bofors to engage them. At 11.00 p m three positions fired a total of 15 rounds and, half an hour later, another position put up a further 12 rounds. No hits were claimed.

During the latter part of August two Australian AA batteries were attached for instruction on the Bofors guns but there was no further action. In fact they had to wait until October 16th before they fired again when four guns put up 103 rounds without claiming hits.

November was busier for the light guns, with actions on the 17th, 19th and 24th during which 92 rounds were fired.

As Wavell prepared for Operation Compass — the attack on the Italians in the Sidi Barrani area — X Troop at Aboukir was ordered to move up to Baggush to join Western Desert Force. They arrived at Baggush on December 5th with their four guns and 400 rounds per gun, and stayed there until December 27th when two guns moved to Sidi Barrani and the other two to Buq Buq. A month later they moved back to Mersa Matruh and returned to Aboukir on January 29th 1941.

At Alexandria there had been a small redeployment on January 22nd with two guns being moved to new positions near the floating dock as an additional defence against possible divebombing attacks.

Seven of the mobile guns were traded for static weapons in March when Predictors were also received and put into position. Incidentally three of the mobiles handed in by 5 LAA Bty were taken over by 6 LAA Bty which had also arrived in the Middle East.

The addition of a Troop from 56th (Australian) LAA Regiment on April 11th increased the Battery's strength to sixteen guns, the Aussies moving into new positions in the harbour area. A second Australian Troop, from 15th LAA Regiment, was added on April 30th and they occupied positions at Amiriya aerodrome. The Battery's strength was now 357 officers and men, of whom 116 were Australians.

There were three actions during May the heaviest being on the first day of the month when 152 rounds were fired in a period of two minutes. The subsequent engagements were much lighter with a combined total of only 15 rounds.

Close association with men from the Antipodes continued, with a cadre from 14th New Zealand LAA Regiment arriving for training on May 27th and remaining until June 12th. The New Zealanders were able to witness action on three occasions in early June with 114 rounds on June 4th and 180 on June 7th.

Next day the Battery diarist claimed four hits on an enemy recce aircraft which subsequently crashed. In fact this was the Cant which was hit by 24 Bty from Ras-el-Tin at 28,400 feet (see Part 1, Chapter 6). Since the crippled plane crossed Agami Point at 26,000 feet before being shot down by fighters this particular claim for 5 LAA Bty must be discounted.

Later in the month the Battery's guns were in action again firing 174 rounds on June 17th, 26 rounds on June 20th and 83 on June 23rd. On each occasion the raiders came over between 4.00 a m and 5.00 a m.

313 rounds were put up by eleven of the guns on July 19th during a barrage defence of the harbour, and a week later six guns fired 37 rounds of barrage fire. Those were the last shots fired by 5 LAA Bty as part of 9th Regiment, as that day — July 26th 1941 — the Battery became part of 52nd LAA Regiment commanded by Lt Colonel C H Mather.

The Ards men did not leave "Alex" immediately but remained there until September. In the meantime they were in action on a number of occasions during August, on one of which, on August 7th, the gun at D Position was put out of action by a bomb which landed nearby. Five men were injured, but fortunately none fatally. Eleven guns fired a total of 842 rounds in barrage fire, the Battery's highest rate of fire to date. A 257-round barrage was put up on August 31st.

Orders were received to move up to Tobruk on September 19th. At the time Tobruk was under siege by Rommel's forces and resupply had to be carried out by sea. Thus the Battery had to leave their guns and equipment in "Alex", handing these over to 22 LAA Bty, Royal Malta Artillery. By September 23rd 200 officers and men of the Battery were at Tobruk, where they relieved 8 (Australian) LAA Bty and came under the command of 14th LAA Regiment. The equipment which they received from the Australians was the Italian Breda 20 mm gun of which they took over eighteen examples.

They had hardly had time to become acclimatised to the conditions at Tobruk when they were caught in an air raid at 10.00 a m on September 25th. Julian Dorrian remembers that the Australians had already warned the Irishmen that a regular recce plane came over each morning, and it seems that this machine spotted an irresistible target in the form of men of 5 LAA Bty drawn up in their camp and whistled up the bombers. In the subsequent onslaught the Battery had six men killed and another twelve injured. The dead were L/Sgt John McNeilly, Gnrs Sammy Graham, Joe Dalzell, Jim Meredith, Charlie Boyes and Paul Robson. The most seriously injured were L/Bdr Willie Pritchard and Gnr Robert McCracken, each of whom lost a leg, while a number of others, including Pritchard's brother Jim, had to be evacuated to hospital with shrapnel wounds.

From then on their time in Tobruk was to be spent largely underground except when they were manning their guns, and they were to share in the nickname which the defenders applied to themselves because of this underground lifestyle — the "Rats of Tobruk".

Providing air defence for field artillery units, they went into action on September 30th firing 11 rounds at a German plane. During October they opened up on enemy aircraft on five occasions, on the last of which they claimed a hit on an attacker for an expenditure of 206 rounds. Of these 189 rounds had come from seven Bredas, while 17 were from two of the three Bofors which they had received on October 19th in exchange for four of the Bredas.

November proved to be an extremely active month with ten engagements against enemy aircraft including fighters and Ju87 dive-bombers (Stukas). Hits were claimed on five occasions although there was no confirmation of damage or destruction. Axis field artillery damaged a Bofors on November 22nd, but the siege of Tobruk was almost over. Even so they were involved in a further five actions before the relief on December 7th. Although the worst was over there was an attack by a force of over twenty Ju87s and a similar number of fighters next day, and on December 9th the Battery shot down a Ju87 from an attacking force of fifteen machines.

5 LAA Bty was then attached as air defence to the Polish Brigade and left Tobruk with the Carpathian Regiment of Artillery on December 12th. Equipped with three Bofors and nine Bredas they moved to El Adem where they were positioned across the Tobruk-El Adem road. From these positions they shot down another enemy aircraft on December 13th.

Operation Crusader was now pushing Rommel back towards Tripolitania, and the Battery moved up through Acroma and Gazala where two men were wounded to Benghazi, leaving three guns at the Derna Pass. They entered Benghazi two days after its fall on Christmas Eve and came under the command of 68th HAA Regiment for harbour defence. They were in action in this role on a number of occasions before the withdrawal from Benghazi after Rommel's counter-attack in January.

CHAPTER SEVEN

DEATH OF A BATTERY

Benghazi changed hands again on January 28th 1942, just seven days after Rommel launched his new offensive from the El Agheila positions. Auchinleck's forces, which had actually been numerically inferior to Rommel's at the beginning of the "Crusader" offensive in November, had been weakened by the diversion of much-needed reinforcements of men and equipment to the Far East. With Rommel's armour reinforced there was no alternative for Eighth Army, faced with this fresh onslaught, but to withdraw to a more tenable position.

5 LAA Bty accordingly were pulled out of Benghazi, having lost another one of their number when Gnr Eric Shiels was killed in an air raid. The Battery pulled back via Derna where it regrouped taking the detachments which had been left there during the advance under command once again. The Ards men were brought back as far as Mersa Matruh where they handed in their Bofors Mark 1s for the newer Mark 2 equipment. This done they prepared to move forward again and found themselves back at Tobruk in March.

Eighth Army had stopped Rommel's advance at the Gazala-Bir Hakeim line some 30 miles west of Tobruk and the two armies faced each other along this line until May 26th when the Desert Fox attacked again. While General Cruwell advanced on the Gazala area with two Italian corps and one German brigade, Rommel outflanked the Bir Hakeim position with an Italian corps and the German *Afrika Korps*.

Rommel's intention was to move north behind the British lines while Cruwell pinned down the defenders at Gazala, but this did not meet with immediate success. Fierce fighting raged at Bir Hakeim, which was held by Free French troops. At the Cauldron, Ritchie, commanding Eighth Army, counter-attacked on June 5th/6th. Ritchie's move was too late, for the Free French were forced out of Bir Hakeim four days later and on June 11th Rommel broke through 30 Corps. June 18th, the anniversary of Waterloo, saw Tobruk surrounded yet again.

On this occasion however there was to be no prolonged siege. The defences which had held Tobruk a year before had been allowed to decay. Minefields had been lifted to be relaid at Gazala, anti-tank ditches had filled up with sand, and there simply were not enough anti-tank guns in the town to provide a defence against any determined attack. Indeed Auchinleck had decided that Tobruk should not be held; the siege of '41 had put too great a strain on his resources and on those of the Royal Navy for very little strategic return. However the siege of Tobruk had had a tremendous effect on the public imagination in Britain and indeed on Churchill himself, and for reasons which were symbolic rather than military the Prime Minister wanted Tobruk held again.

Whatever Churchill's wishes there was simply no time to organise a defence of the town, which was doomed from the moment Axis forces, in the shape of three armoured divisions, moved against it on the morning of June 20th. In the confused fighting which followed 5 LAA Bty were engaged as AA and anti-tank gunners and it is impossible to give a completely accurate account of all that happened to the gunners as Tobruk's situation became even more desperate. A selection of memories of those grim hours does however convey the feel of what it must have been like.

John Atcheson was on Sgt Adair's gun located on the Derna road where Gnrs Bobby Bailey and Arthur Gage were killed before the position was overrun and the remainder of the crew including Atcheson were captured.

Near the NAAFI bulk store Sgt Alec Davidson's gun engaged the enemy until it was put out of action by enemy fire which killed Davidson himself as well as Captain John Gibson and Gnrs David McClinton and Rayner.

Sgt Jack McFadden's gun, on top of a wadi, was more fortunate. None of the crew were killed and some tried to escape. One man succeeded — Julian Dorrian got out through the harbour before it fell that evening.

The fighting continued into the next day when the last Allied troops laid down their arms in the evening. Probably the last gun of 5 LAA Bty to fall was that of L/Bdr George Pollock under Lt Carver (who had been recruited from the British community in Egypt). Pollock's gun was engaging Ju87 dive-bombers when it jammed; he had just cleared it when German infantry surrounded the crew and took the entire detachment prisoner.

As well as Julian Dorrian one other man escaped. John McGinn had just been evacuated by ship for hospital treatment hours before the attack struck. Both were later drafted to 6 LAA Bty, showing that the Base Depot could have a little heart at times.

For the remainder of the Battery it was however a PoW cage — except that is for the dead who rest in the British Military Cemetery to this day. As well as those whose deaths are described above Gnrs Bobbie Cromie, John Miskelly and Edward Flynn had also been killed in action while Harold Johnston had been killed by a defective shell in his own gun position before the attack.

5 LAA Bty had thus ceased to exist and for most of its men the remaining three years of the war were to be spent as prisoners. Some escaped and were recaptured like John Atcheson and Jim Pritchard. John Atcheson escaped in Italy and lived for a time with an Italian family on their farm before he was betrayed by Fascists. He was sent to Germany by train, but at Florence the train was hit by Allied bombers and Atcheson escaped again from the chaos in which some 700 men died. His luck did not hold out for he was later recaptured once more and spent the remainder of the war in captivity.

Jim Pritchard was on the run for three months before he was recaptured and remained a PoW until liberation by the advancing Allied armies. Tom McCutcheon however died in a PoW Camp, while Willie McBlain was killed in an RAF air attack while he was a PoW.

The fall of Tobruk was seen as a tremendous triumph for the Germans and Italians and an unmitigated disaster for the Allies, especially Britain. Churchill had to face a vote of "no confidence" in the Commons while Rommel was promoted to Field Marshal by Hitler and thus achieved the fastest rise ever to this rank in the German army. Hitler's decision to promote Rommel is a firm indicator of just how much the defence of Tobruk had rankled the *Führer* in 1941. Rommel was also given permission to push on towards Egypt, an undertaking which, although at first appearing to be another glittering success, was to lead the Desert Fox to ultimate defeat in North Africa.

In his book "With Rommel in the Desert" Heinz Werner Schmidt expresses the view that Tobruk fell because no contingency plan had been made to meet a headlong assault and because most of the armour needed to defend the town had been destroyed in the desert battles around Gazala and the Cauldron. Tobruk had in fact contained a fresh South African Division, but this formation had been positioned to meet an attack from the south-west and Ritchie had, contrary to Auchinleck's orders, withdrawn Eighth Army and had thus left Tobruk hopelessly exposed. The inexperienced South African commanders decided to surrender rather than lose "the flower of South African manhood". That such a decision was not appreciated by that "flower" was ably demonstrated by the bitter reaction of the South African soldiers to the sight of the white flag on their headquarters.

The shame felt by the South African soldiers need not be shared by the men of 5 LAA Bty, for the Ards men fought well and valiantly in the struggle for Tobruk.

No one could have asked more of them for they went, in the words of the Royal Regiment of Artillery's motto, where duty led.

AUTHOR'S NOTE

No War Diary has survived for 5 LAA Battery for the period January to June 1942. The Battery's history for this period has had to be constructed entirely from interviews with a small number of survivors of those days; it therefore lacks the detail that other sections of this book include. In addition the Battery's War Diary for 1940 and 1941 is not as detailed as those of other units; for example, the names of those killed at Tobruk in September 1941 are not listed.

PART SEVEN

APPENDICES

APPENDIX ONE

THE 3·7 INCH HEAVY ANTI-AIRCRAFT GUN

At the end of the First World War the standard British anti-aircraft gun was the 3 inch 20 cwt weapon which had been designed under Royal Navy auspices prior to the war and had been accepted for both army and navy use. This gun continued in use in the post-war years, when it was also planned to develop a 4·7 inch gun for the defence of rear areas and of such points as naval bases. In 1928 a 3·7 inch gun firing a 25 lb shell to a ceiling of approximately 28,000 feet was mooted as effective to "fill the gap between the medium 4·7 inch and the 3 inch weapons". Five years later a specification was issued for a 3·7 inch. This weapon which would weigh 8 tons had to be capable of being brought into action in 15 minutes. It was also to be mobile and a speed on tow of 25 m p h was called for.

Thus was born the 3·7 inch gun with which 9th Regiment's heavy batteries fought. The design accepted for production came from Vickers-Armstrong and the prototype appeared for trials in 1936. The trials were successful and in April 1937 production orders were placed, with the first guns being delivered in January 1938. In its production form the gun exceeded the weight specified by more than a ton. However it also exceeded specification in being able to fire a 28 lb shell to a height of 41,000 feet at the rate of 10 rounds per minute. Although this was an "ideal" situation there was no doubt that the 3·7 was an excellent gun, and this was to be proved, again and again, in the years ahead.

That height of 41,000 feet needs some explaining. This was in fact the maximum "ceiling" to which the gun could project a shell but what mattered in service was the "effective ceiling" which was governed by the maximum elevation of the gun and the operating time of the fuze fitted in the shell. Trials guns could fire fuze-less solid rounds whilst pointing vertically upwards, but guns in action had to fire explosive-carrying shells for which the maximum fuze-operating time in 1938 was 25 seconds. Since the gun's normal maximum elevation was 80°, this, and the fuze time, reduced maximum ceiling in action to 28,000 feet. Reduction in the gun's elevation caused by the movement of the target could reduce the ceiling even more. When the 3·7 entered service the effective ceiling, defined as that height at which a gun could engage an approaching target which was flying at 250 m p h for 20 seconds before the elevation came down to 70°, was described as 23,000 feet. However, during the war as fire control improved and new and better fuzes were introduced, the effective ceiling of the 3·7 rose to 32,000 feet for the Marks 1 to 3 and to 45,000 feet for the later Mark 6. Maximum ceilings for these weapons were respectively 41,000 feet and 59,300 feet.

To the gunners of 1938 the 3·7 was an extremely advanced, complicated and very heavy piece of ordnance. Because of this it was initially not liked by the men, who were used to the much lighter and easier to handle 3 inch gun. Indeed the 3 inch gun was chosen to accompany the BEF to France in 1939, leaving the 3·7 for home defence. However some 3·7s were sent overseas and one of the bases which received these guns was Alexandria. The first guns supplied to Alexandria were manned by 20 AA Bty who handed some over to 9th Regiment's batteries as their initial equipment. By May 1940 the Regiment had received its entire complement of 24 guns.

Initial lack of enthusiasm for the 3·7 soon gave way to appreciation of the excellence of the gun which at the time of its introduction was undoubtedly the best of its kind in the world. Even six years later the 3·7 was still rated as one of the best of its type in existence anywhere.

In its Mark 1 form the 3·7 was a mobile gun fitted with magslip dials for the display of data relayed from a predictor by electrical transmission, thus allowing the gun-layers to "line-up" the gun's elevation and bearing on the target. Initially open sights were also fitted as was a loading-tray. The loading-tray was later removed but was restored when automatic fuze-setters were introduced.

For the defence of fixed points a static version was also produced and this was designated Mark 2. This variant was bolted to a holdfast, although it could be transported on a special limber. One of the most obvious differences, after the absence of a carriage, between the Marks 1 and 2 was the counterbalance fitted to the static gun rather than the spring balancing system of the mobile gun. Gunlayers on the static gun faced rearwards rather than forwards although open sights were fitted initially to this gun also. With open sights removed and some other minor variations the Mark 1 became the Mark 1A. A lighter, less complicated wartime production gun-carriage produced what was known as the Mark 3. [Although the gunners used these designations to refer to the entire weapon they really refer to the mounting rather than to the gun itself which also went through a series of mark numbers. Throughout the text however I have used the terms which were accepted at the time and thus the Mark 3 guns mentioned in the text are actually guns mounted on Mark 3 mobile carriages.]

All of the early versions of the 3·7 were capable of firing a 28 lb high-explosive shell using a charge of 8½ lbs of cordite at a muzzle velocity of 2,600 feet per second to a maximum ceiling of 41,000 feet (see above) or for a horizontal distance of 20,600 yards. Hand-loading gave a rate of fire of 10 rounds per minute from a good crew although during the Battle of Tripoli some of 9th Regiment's crews often exceeded this rate. Automatic loading increased the firing rate to 25 rounds per minute, but this did not come until later in the war. The gun itself had a bore-length of 185 inches which was fifty times the calibre of 3·7 inches and was thus described as being 50 calibres long in ordnance jargon.

The later Mark 6 gun was in fact a 4·5 inch gun fitted with a barrel-liner to reduce its calibre to 3·7. The bore-length of this piece was 65 calibres and although the maximum ceiling was almost 60,000 feet and the effective ceiling 45,000 feet the rate of fire was reduced to 8 rounds per minute by hand-loading and 19 by auto-loading.

"Long-hand" was a post-war development of the 3·7 which, with automatic ammunition feed, was capable of firing 60 rounds per minute. It was however overtaken by the guided missile and never entered service. The 3·7 itself finally gave way to the Thunderbird surface to-air missile (SAM) after more than twenty years in service.

Even towards the end of the First World War the increasing speed of aircraft had begun to create difficulties for anti-aircraft gunners. To fire at the exact point in the sky which the aircraft filled was to fire at a target that was no longer there! It therefore became necessary to fire in front of the aircraft so that by the time the shell reached the aircraft's altitude the two would, hopefully, coincide and the plane would be shot down. Dead-reckoning or inspired guesswork could hardly be expected to be adequate in this task and this led to the demand for equipment which could predict the future position of an attacking aircraft, given its existing bearing, height and speed. At the same time as the 3·7 was being designed the Predictor which would prove so necessary was also being developed. Today a matchbox-sized computer could probably do the job — in 1939 the answers came from Vickers in the UK and Sperry in the USA in the shape of Predictors which could weigh up to 1½ tons and contain about 3,500 parts!

Basically the Predictor was a computer which, when fed with the present position, speed, height and course of an aircraft together with data such as windspeed and direction plus the gun's position, could calculate the data needed for fire control and tell the gun crew where in the sky they could expect their target to be at the end of the shell's flight. In theory this would allow the shell and its target to meet but of course an attacker was not going to behave with mathematical logic every time. The information was fed from the Predictor to the gun by electrical transmission where dials showed the bearing, elevation and fuze-time needed. Each of these pieces of data went to a separate crew member — two gunlayers and the fuze-setter. One gunlayer (the No 2) laid "for line", i.e., he traversed the gun towards the target (the 3·7 could traverse through 360°) in the horizontal plane, while the other gunlayer (the No 3) laid "for elevation" by raising the barrel to the angle of elevation required. The fuze-setter (the No 4) set, or "cut" the time on the fuze manually. For the gun-layers the system was simple; they each faced a dial fitted with two pointers one of which was set by the Predictor. The layers then swung or elevated the gun until the second pointer "blacked out" the first, thus ensuring that the gun was on target.

To enable the Predictor to do its job it had to be provided with the relevant information which required a number of other personnel in addition to the Predictor crew. Height and range were determined using a Height and Rangefinder, a massive piece of equipment which looked like nothing more than a huge metal tube mounted horizontally on a swivel stand. The tube was fitted at either end with sights and the operator observed through two eyepieces the images projected in front of him. By adjusting the instrument he could bring the two images together and when this occurred target data could be read off a scale and passed to the Predictor — it was normal to take three readings and to use the mean of these for feeding into the Predictor.

A factory-fresh 3·7 inch gun on its mobile carriage. Note the massive jack-arms which swung up for transport; these dropped to provide a steady gun platform in action.

(Photo: Imperial War Museum)

A Predictor crew at work with a static 3·7 in the background.

(Photo: Imperial War Museum)

Spotters identified incoming aircraft as friendly or hostile and passed the bearing to the Heightfinder. This job was eventually made much simpler by the introduction of radar, although the spotter was not made redundant as the friend or foe identification still had to be confirmed.

Originally known as radio direction finding (RDF), radar provided early warning of the approach of aircraft. Friendly machines could be identified by the use of a device called IFF (Identification, Friend or Foe) which transmitted an identification signal to the defences and the approach of incoming hostile aircraft could therefore be monitored and the defences alerted. Of course IFF could break down, or be damaged in action, or the pilot could simply forget to switch it on, so close-in checking of the identity of aircraft was always important.

Attempts had been made to pick-up approaching aircraft with various sound-locating devices, some of which in the shape of huge concrete dishes, survive to this day, but these were rapidly made obsolete by the introduction of radar. Radar was quickly refined to the extent that individual aircraft could be tracked and the Gun Laying (GL) radars came into use. As GL radar improved throughout the war it greatly simplified the task of the gunners. Linked with the vastly improved Predictor No 10, the excellent American GL radar known as SCR 584 took much of the labour out of the anti-aircraft gunner's job. The introduction of Remote Power Control (RPC), added to the automatic setting of fuzes, all allowed the guns to be fired almost automatically.

One of the drawbacks to the 3·7 (or any other anti-aircraft weapon) was the need to set fuzes by hand. It was in this process that most error could creep in and this was especially so during a long engagement with sustained high rates of fire when the No 4 was tiring, allowing the "dead time" between the "cutting" or setting of the fuze and the firing to increase. However the Molins Machinery Company, who in peacetime produced cigarette-making machinery, turned their attention to war-work and, in addition to producing the famed Molins gun, devised the Machine Fuze Setter No 11 (MFS 11) which not only set the fuze, giving a constant "dead time", but also operated the loading mechanism on a 3·7. The loading tray was now restored to the gun and when a round was placed in the tray the MFS 11 engaged it and set the fuze before withdrawing whereupon the tray swung over to line up with the breech and begin the firing operation. This meant that the gun could now fire up to 25 rounds per minute rather than the 10 per minute with manual fuze-setting and loading.

Fuzes themselves were also the subject of much wartime development. Before the war fuzes were bought from Switzerland and when that source of supply became unavailable British industry tried desperately to perfect good clockwork fuzes. Their success was not however 100% and many "dud" fuzes found their way to guns, some of which exploded prematurely, killing and injuring gun crew members.

While industry struggled with the clockwork fuze the Pye company was working on an entirely different type of fuze — the proximity fuze. This had been born out of the application of radar to gun control, when it was believed that the logical development would be to fit some sort of radar into the shell to detect the energy being reflected from the target. The technology for this approach did not however exist in the 'forties, and thinking turned instead to fitting a miniature radio transmitter and receiver inside the shell so that it could transmit its own signal and explode on receiving a "return" from its target.

A similar principle, using a photoelectric cell had been used in rocket fuzes since 1940 and all of the theoretical work on a radio-based proximity fuze was handed over to the Americans later that year. What came to be known as the VT Fuze (Section V of the US Navy Bureau of Ordnance's Project T) was perfected and went into use in Anti-Aircraft Command in 1944. Later that year it was introduced into field artillery, proving extremely effective in the "Battle of the Bulge". The VT Fuze was of tremendous value against the V1 in 1944 and 1945 and its use obviated the need for fuze-setting of any kind. This "thinking shell" was a harbinger of what was to come in future years as it spelt the end of the gun as the projector of anti-aircraft weapons.

APPENDIX TWO

SACRED HEART CONVENT, ALEXANDRIA

Trudging into Sidi Bishr that November afternoon in 1939 many men wondered when they would ever see home again. Here they were, almost 3,000 miles away from Ireland, in an alien environment and an alien culture. In the days following their arrival they began to become accustomed to the continual cries of begging Arabs — and they themselves were not much better off — and began to develop an Arabic vocabulary. Off the main streets of Alexandria the smell of garlic mixed with human sweat was all-pervading, though it was nowhere near as bad in "Alex" as it was in Port Said, as many found when they went to gun positions there. The sound of an Irish voice in this strange land would be music to their ears.

The music came in a way that few would have expected for it was not long before some of the "Derry Boys" discovered that just down the road from Sidi Bishr in the Ibrahamieh district there was a school which was run by nuns of the Franciscan Order — and most of the nuns in the adjoining convent were Irish!

In fact one of those sisters actually came from Derry. Sister Mary Richard Coyle was a member of a very well-known family from Donegal Street in Rosemount and she had been in Alexandria since shortly after her profession in Rome in April 1939. Sister Richard, who today is in a convent in New Jersey, USA, recalls that some of the gunners came along to the convent rather diffidently at first but on discovering that they would be made welcome returned with even more of their comrades.

Rules for the religious orders were much stricter in those days than is the case today and although the sisters could provide the visiting gunners with tea and sandwiches they could not eat with them. Even so Sacred Heart Convent and its adjoining school soon became familiar territory to many of the men of Ninth Regiment irrespective of their religious denomination.

Of the fourteen sisters in the Convent on Rue Tanis no fewer than nine came from Ireland and their primary function in Alexandria was to run the school which

Captain Hume Stewart-Moore of 26 HAA Battery who commanded MAC Troop for much of its existence.
(Photo: Sir Basil McFarland's Collection)

Gnrs J J Doherty and Denis Farrell outside Sacred Heart Convent. Both men served in 25 HAA Battery and both came from the same townland of Drumenny near Strabane in County Tyrone.
(Photo: Author's Collection)

adjoined the Convent, although it opened on to the Corniche. The school itself was put at the disposal of the soldiers for educational and social purposes. A number of gunners attended maths classes organised by Sister Richard and they proved to be most enthusiastic pupils indeed. The school hall was turned over as a social club for the soldiers on at least one evening per week when they could gather for a peaceful evening and enjoy tea and sandwiches and perhaps have a sing-song. The bread and fillings for the sandwiches were provided by the men themselves, and the army bakers produced extra bread for the Sisters which they found preferable to the inferior fare available in Alex. The ingredients for an evening's sandwiches were left at the convent during the afternoon and the Sisters would prepare them after they had finished their own evening meal. Some of the soldiers arriving at the school for the evening would collect the sandwiches from the Convent kitchen and take them through to the school hall. None of the Sisters could join in those evenings as the Order's rules in those days imposed a silence on them from evening until morning.

The visitors to Sacred Heart Convent were not only Irish; there were also English, Welsh, Australian and New Zealand troops along with South Africans — whites and Zulu and Basuto soldiers — although Sister Richard remembers very few Scots.

J J Doherty was one of the most frequent visitors to the Convent and he spent many off-duty hours there. In fact J J became the Convent's unofficial handyman after the Sisters discovered his skill at woodwork and at carrying out repairs on anything from the plumbing to the lighting.

During Ninth's Regiment's time in Alex the "Derry Boys" came to the aid of the Sisters in an unusual fashion — by writing letters home for them. As army mail had more chance of reaching the UK than civilian mail (and since the Order restricted communication anyway) the soldiers offered to write to the families of various Sisters and pass on news for them. J J Doherty was one of several who wrote on behalf of Sister Mary Richard to her family in Rosemount (incidentally this spate of letter writing confirmed to the folks at home the exact location of their own Regiment). Sister Richard's older sister Anna offered to answer J J's letters because she said that his writing was the most beautiful she had ever seen. Soon J J and Anna Coyle were corresponding with each other and they kept up this penfriendship until the Regiment returned home on leave in 1944, when they were married in St Eugene's Cathedral. (Their first son was named for both his father and the aunt who had brought his parents together.)

The Franciscan Sisters remained in Sacred Heart Convent throughout the war except for a brief period in the summer of 1942 when, due to the fear of a German/ Italian breakthrough to the Delta, they were evacuated to Palestine. After this they returned to "Alex" in time for the new school year and for the opening of the October Battle of El Alamein — when the Convent shook to the pounding of artillery sixty miles away!

Before the evacuation two young nuns were due to take their final vows on July 2nd and the profession ceremony was to be conducted by a Forces chaplain who had to inform the Mother Superior that he might not be able to attend due to the war situation. In the event he did conduct the ceremony and was afterwards able to give the Sisters some information about a large explosion which they had heard the night before — it had been caused by a Ju88 dive-bombing the harbour area and scoring a direct hit on an ack-ack gun pit! Fortunately the guns and the gunners had been evacuated as Alexandria had been declared an "open" city and thus no one was injured.

Sister Richard also recalls the many air raids which occurred, causing the Sisters to spend the night in a shelter in the school grounds where they prayed for the people underneath the bombers, for the ack-ack gunners and for the airmen. If however they had not left the Convent by the time the guns opened fire they had to remain indoors due to the amount of debris falling from the shells!

In 1947 when Sister Richard visited home for the first time in more than a decade one of her first visitors was an ex-sergeant of 24 Battery who told her that his dying mother would like to meet her. Sister went to the home of the dying woman who said that she had prayed to live long enough to thank Sister Richard for her kindness to her son and all his comrades in Alexandria. When her soldier son had written home to say that he was visiting a convent his mother had felt secure for his well-being — both physical and spiritual. Forty years later Sister Richard still remembers clearly the sincerity of that dying lady and the decency of her son whom Sister never saw again, for he died before her next visit home. There had never been any religious qualification for visitors to Sacred Heart Convent and that particular soldier had been one of many Protestant visitors.

Hospitality has for centuries been a tradition of the religious orders. The hospitality shown by the Franciscan Sisters in Alexandria to lonely soldiers so far from home did much to sustain those men as they came to grips with their new situation and surroundings. It was truly appreciated by all of them and who can doubt that the continuing prayers of the Sisters contributed to the fact that Ninth Regiment sustained comparatively light casualties throughout the war.

APPENDIX THREE

MAC TROOP

The demonstration of 3·7s in the field role at Azizia in February 1943 so impressed Brigadier Weir, the New Zealand CRA and General Freyberg that the New Zealanders requested that HAA weapons be placed at their disposal for the fighting in Tunisia. That request was acted upon by the authorities who provided four such guns for a special troop to go forward in support of the New Zealanders. However the guns which were used to equip this unit were not 3·7s but captured German weapons — the redoubtable 88 mm guns of which a number had been captured during the advance on Tripoli. A number of Italian 90 mm weapons were also taken into service for AA purposes in Tripoli.

During the Desert campaign the 88 had established a fearsome reputation as an anti-tank and field gun although their only real advantage over the 3·7 in these roles lay in the fact that the 88 was a lighter piece of ordnance.

As the guns were under the control of 2nd AA Brigade they were formed into an *ad hoc* troop composed of personnel from that brigade and named after the Brigade Commander, Brigadier H M J McIntyre, who was known as "Mad Mac". Thus the Troop was dubbed MAC Troop. Its personnel came from the HAA regiments in 2nd AA Brigade, with 36 men coming from 9th HAA Regiment.

On March 5th 1943 MAC Troop arrived at the rear of 2nd New Zealand Division and was placed under the command of the CRA of the Division; the troop itself was commanded by Captain Downing of 51st HAA Regiment, with Lt Hume Stewart-Moore of 26 HAA Battery as the second-in-command.

The Battle of Medenine was in full sway and positions were reconnoitred for MAC Troop's deployment by Lieutenant Colonel Sprosen. The gunners of MAC Troop first went into action with their captured guns against a pair of Ju87s which attacked the NZ Artillery HQ from high-level. So quickly did they go into action and fire however that they forgot to set the time-fuzes and several rounds fell back to earth in a NZ Army Service Corps area, exploding as they hit the ground.

MAC Troop then moved into position near Metameur coming under the direct control of 4th (New Zealand) Field Regiment with whom they would remain until the end of the North African campaign.

The Troop fired in the field role with some diversion to anti-aircraft fire during the Battle of Medenine and then moved up to the Tebaga Gap for the assault on Point 201 on March 20th and 21st. MAC Troop suffered nine casualties here as a result of German counter-battery fire, which was very effective and even managed to bracket the Divisional Artillery HQ's Armoured Command Vehicle as it moved forward. In the subsequent attack by New Zealand infantry the guns of MAC Troop participated in the supporting artillery barrages.

As they moved up towards Wadi Akarit from Mareth, having come through the breaking of the Mareth Line, MAC Troop suffered its most severe casualties to date. On March 27th, during the advance, a Bantam car (jeep) carrying Bombardier Eric (The Judge) Rutherford and Gunner John Kerr was blown up by a landmine. Kerr suffered a broken right arm and other, less serious, injuries but Rutherford was so badly injured that he died in the field hospital. Both men were members of 26 Battery, as was Gunner Fred Barrie who lost an arm as a result of enemy action in a separate incident.

On March 30th MAC Troop was in position for the forthcoming attack on Wadi Akarit joining in the artillery programme and again coming under counter-battery fire. They also engaged a number of Ju88 bombers which attacked the positions during the afternoon and evening.

The Battle of Wadi Akarit opened on April 6th with an artillery barrage which, in terms of concentration of fire, surpassed that at El Alamein. A total of 496 guns from twenty-two regiments (including the four guns of MAC Troop) poured a storm of fire on to a German front of about twelve miles.

As the New Zealand Division were in reserve their guns supported 50th Division and fired in a phased programme from 4.15 a m to 9.41 a m loosing off more than 200 rounds per gun.

There was very little counter-battery fire but enemy aircraft did attack the British troops and MAC Troop came under strafing and bombing attack. One *Luftwaffe* fighter-bomber dropped a number of bombs on MAC Troop's positions scoring a direct hit on one gun. The explosion killed one of the crew instantly and another died minutes later as one of his comrades tried in vain to help him. Three other men were injured. Both of the dead were 25 Battery men from Derry, Bombardier Neil Murray, a former member of the Irish Army, and Gunner Tommy McCarter. A New Zealand medical orderly, Bombardier L H Shearer, rushed to the aid of the injured even as bombs continued to fall in the area and earned the highest praise from the gunners.

After a short period in a 'laager' area which was infested with mosquitoes and from which working parties were sent out to prepare new positions 6,000 yards to the south-west of Takrouna 4th Field moved into position for the attack on Takrouna on April 19th. During this battle the regiment had the benefit of an Air Observation Post (AOP) and MAC Troop again took part in the artillery programme. This was in spite of the fact that the barrels of their 88s were so badly worn that the guns could no longer be ranged accurately.

Their final engagements were on May 11th and 12th. Supporting Free French forces they contributed to the heaviest ever counter-battery artillery programme of the war in North Africa. German counterfire was generally ineffective even though it included airbursts and MAC Troop were able, literally, to give the German artillery a taste of their own medicine by showering them with airbursts from the 88s.

Next day the German artillery used a new, heavy calibre (210 mm) long range rocket projector for what appears to have been the first time. 4th Field Regiment's RHQ was hit by rockets causing considerable damage but fortunately there were no injuries. Other units were not so lucky; when RHQ of 111th Field Regiment, RA was hit the CO was one of a number who were killed. The rockets were being fired from a gully and it was no easy task for the 25 pounders of 4th Field to engage them. However MAC Troop were able to do so and fired airbursts over the troublesome rocket-launchers, silencing them effectively.

During the course of that afternoon German troops had begun to surrender and by the end of the day von Arnim had surrendered his Army Group Africa. Over 100,000 German prisoners were taken and to this could be added almost 90,000 Italians. Other Axis troops brought the total to over 230,000.

With the war in North Africa over MAC Troop was disbanded and its personnel returned to their parent units — Hume Stewart-Moore rejoining 26 HAA Bty as a captain, as he had replaced Downing as Troop Commander. In its brief but hectic existence it had performed an excellent job and brought credit to the Regiments which had provided its manpower (S/Sgt Elmour Fallis of 26 Bty being highly commended for his work with the Troop), not least 9th (Londonderry) Regiment for it was the "Derry Boys" who, unfortunately, made up the bulk of MAC Troop's losses.

APPENDIX FOUR

THE FATHER OF THE REGIMENT: A TRIBUTE TO SIR BASIL ALEXANDER TALBOT McFARLAND Bt CBE ERD ADC

"The Father of the Regiment" was how one fellow-officer of the Ninth described Sir Basil McFarland, Bart. It is a tribute that sums up the feelings which the men of the Regiment held for a remarkable officer, and the truth in that tribute was demonstrated when the War Office appointed him as Honorary Colonel of 9th Regiment in 1946.

Basil Alexander Talbot McFarland was born on February 18th, 1898, the only son of John and Annie McFarland. His father was created 1st Baronet Aberfoyle in 1914 and on Sir John's death in 1926 the young Basil succeeded to the title.

His education had been cosmopolitan and embraced schools in three countries — Germany, Belgium and England. While still a teenager he joined the army, serving as a private in the Artists' Rifles (28th Battalion, The London Regt) in the final year of the Great War.

After the war he returned home to Londonderry where he took an active part in the social, sporting and business life of his city. He played rugby for City of Derry and NIFC and was an excellent player — so good in fact that he was selected for Ireland on four occasions between 1920 and 1922. He later took an active interest in the newly-formed Derry City Football Club and was a director of the club for a number of years.

In 1924 he married Annie Kathleen Henderson of Whiteabbey and their marriage was blessed with a son and a daughter. His standing in his native city was all the while increasing in every sphere and in 1930 he was appointed High Sheriff, an office he was to hold until 1939. At this time he was also a member of Londonderry County Borough Council and in 1939 the Councillors elected him to the Mayoralty — an office which his father had also held.

At the time of his election as Mayor he had also been commissioned into the 9th Anti-Aircraft Regiment, Royal Artillery (Supplementary Reserve), being promoted from Lieutenant to Captain in August 1939. When war was declared he resigned the Mayoralty and, although more than eighteen months past his fortieth birthday and one of the oldest members of the regiment, prepared to take his leave of the city for whatever the war held in store for him.

As Battery Captain (second-in-command) of 25 Battery he established the practice of paying the soldiers personally at the weekly pay parades and soon came to know every member of the Battery. He acceded to command of "Two-Five" in November 1941 and was promoted to major shortly afterwards. He continued to maintain close contact with his gunners, even taking the pay parade himself from time to time. His men found that Sir Basil always had their welfare at heart and he was prepared to circumvent bureaucracy for the good of the Battery. He never forgot to express his thanks when a job was well done and as a result almost never had to criticise a job that was badly done. The gunners responded by showing Sir Basil an affection and loyalty that could hardly have been bettered. On December 16th 1940, during heavy air raids on Port Sudan, Sir Basil drove up to each gun position in turn. Climbing up on his car, oblivious to the bombs and falling shrapnel (some of which actually damaged his vehicle) he called to the guncrews "Never forget you're Derrymen"! From that day onwards they loved him!

Social divisions between officers and men went by the board on occasions such as St Patrick's Day and Christmas Day when he would join in the festivities with relish. He was always quick to thank those at home who contributed to the Regiment's Comforts Funds, thus allowing special meals to be provided for the "big" occasions but it is not generally known that some extremely generous contributions were made to the Funds from Sir Basil's own pocket.

His name became synonomous with 25 Battery, so much so that on one occasion when 25 put up a performance that was exceptional even by their high standards the Brigade Commander simply remarked "I never expected anything else from McFarland's Battery".

On two occasions he had the opportunity to return to the UK but turned these down as it would have meant leaving his Battery and the soldiers with whom he had shared both good times and bad in Alexandria, the Sudan, the Western Desert, Tripoli and Italy. In sharing their dangers and in giving inspiring leadership he earned a Mention in Despatches for his conduct at Salerno.

When the Regiment returned home on leave in September 1944 the city showed its appreciation of Sir Basil's service by making him an Honorary Freeman of Londonderry, and few would argue with the statement that his name is one of the most deserving of that distinction in the long roll of Freemen.

April 1945 saw him return to civilian life and he was soon immersed again to the full in the life of the city, taking up his business interests and becoming Mayor once again, which office he held until 1950.

But he had not left the military life behind entirely, for in 1946 he was offered, and accepted, the post of Honorary Colonel of 9th (Londonderry) HAA Regiment. It was a most appropriate, one might say inspired, choice for no one had done as much as Sir Basil to keep the Regiment's identity alive during the war. He had fought to have his own men returned to their Battery after illness or injury in spite of all the worst efforts of the Royal Artillery Base Depots, and when transfers were unavoidable he had done his utmost to ensure that founder members of the Regiment were retained as long as possible. It was Sir Basil who "engineered" the transfer of men from the Ninth to 6 LAA Battery in the Tripoli area in April 1943; the men who went to 6 Light were all "originals" and of course 6 Light had been raised as part of 9th Regiment. And it was Sir Basil who was responsible for the nickname "Derry Boys" from his telegrams home to the city fathers with the famous phrase "all Derry Boys with me are well".

When 246 (Derry) (Mixed) HAA Regiment RA, TA was formed in 1947 he was a natural choice as Honorary Colonel — and one of his officers was his own son. He later was Colonel of 245 (Ulster) LAA Regiment RA, TA. Sir Basil was also Chairman of the Co Londonderry Territorial and Auxiliary Forces Association from 1947

A smiling Captain Sir Basil McFarland photographed in Egypt in 1940.
(Photo: Sir Basil McFarland's Collection)

to 1962 and President of the Northern Ireland Territorial and Auxiliary Volunteer Reserve Association from 1968 to 1971. In addition he had found time to command the City of Londonderry Battalion, Home Guard from 1952 to 1956 and to help form and serve as chairman and president both the Londonderry Branch of the Royal Artillery Association and the Londonderry Services Club.

Sir Basil also held the post of HM Lieutenant for the city from 1939 to 1975 during which time he welcomed two visits by a reigning monarch to the city. He was an Aide-de-Camp (ADC) to King George VI and to the present Queen. In 1954 he was made a Companion of the Order of the British Empire and was also awarded the Emergency Reserve Decoration.

Lady McFarland died in 1952 and Sir Basil remarried in 1955. The second Lady McFarland, the former Miss Mary Eleanor Dougan, died in 1973.

His many business interests included a directorship of the Belfast Bank (1930-1970), and both the Londonderry Gaslight Company and the Londonderry and Lough Swilly Railway Company, and he was for many years a member and Chairman of the Londonderry Port and Harbour Commissioners. His position as a leading businessman enabled him to ensure that no old soldier of the Ninth went without employment after the war. His concern for the welfare of his "Derry Boys" extended to the families of ex-soldiers who died while still comparatively young. A number of charity organisations benefitted from his expertise and generosity, such as the Royal National Lifeboat Institution and St John Ambulance.

When Sir Basil died on March 5th 1986 his beloved city lost one of its most distinguished sons and an important chapter in its history came to an end.

APPENDIX 5

THE TRIPOLI DEFENCE — A SUMMARY OF 9th's CONTRIBUTION, JANUARY TO APRIL 1943

Date and Time	Rounds fired by 9th (Londonderry) HAA Regiment 24 Bty	25 Bty	26 Bty	Total HAA Rounds fired (all units)	No of a/c involved	Comments
25th Jan						
1330	—	—	—	9	1 x Ju88	Engaged by one site only
27th Jan						
0230	—	—	—	0	1	1 bomb dropped. Not engaged
0630	51	16	—	182	1	Minelaying a/c
2nd Feb						
0700-0707	39	43	38	Not recorded	2 x tnk	Bombs dropped
1215	—	—	5	Not recorded	1 x Ju88	Recce
1937-2040	368	392	385	Not recorded	6	Low level bombing attack. No casualties or damage
3rd Feb						
0630	—	16	—	Not recorded	1	Identified as friendly; ceased fire
0720-0730	29	—	37	Not recorded	2	Identified as friendly; ceased fire
4th Feb						
1246	24	28	33	207	1 x He111	Recce at 28,000 feet
2125-2130	32	24	35	Not recorded	1	A/c turned away
5th Feb						
0638-0700	303	288	248	Not recorded	7+	1 a/c only bombed town and outside harbour
0940	—	10	16	52	1 x Ju88 or Me210	Recce
1508	—	—	2	Not recorded	1	Recce at 24,000 feet
1908-1952	188	95	73	Not recorded	5+	Bombs dropped 3,000 yds NE of harbour
6th Feb						
1345	19	—	7	Not recorded	1 x Ju88 or Me210	Recce at 24,000 feet
7th Feb						
1935-2017	251	209	239	1,665	3	1 a/c (2 a/c dam claimed by 26 Bty)
8th Feb						
1106	—	—	—	16	1 x He111 or Me210	Damaged. Prob claimed by RAF
1925-2033	137	42	58	686	7+	1 a/c claimed destroyed by 24 Bty 1 a/c claimed destroyed by 25 Bty 1 allowed as prob. by RHQ
11th Feb						
0630-0738	58	—	—	223	3+	Bombs dropped but no damage
13th Feb						
2000	—	35	—	Not recorded	1	Harbour barrage
14th Feb						
0703	19	62	100	447	2	1 claimed as prob by AA
1959-2025	139	—	71	678	3	A/c turned away by barrage

Date and Time	Rounds fired by 9th (Londonderry) HAA Regiment 24 Bty	25 Bty	26 Bty	Total HAA Rounds fired (all units)	No of a/c involved	Comments
15th Feb						
1500	19	—	—	Not recorded	1	Recce — a/c turned away
1632	2	—	—	Not recorded	1	Recce — approached from sun
2054	9	—	—	Not recorded	1	Engagement broken off on approach of RAF a/c
16th Feb	—	3	—	Not recorded	1	Recce (rds fired as indicators for RAF a/c)
19th Feb						
0515	131	*	136	1,196	6+	Bomb fell on harbour front
20th Feb						
1145	17	11	—	Not recorded	1	Monty visiting Regt at time of raid
21st Feb						
2036-2100	256	*	262	2,480	8+	1 a/c destroyed by HAA
22nd Feb						
0520-0608	281	649	351	3,163	23	3 a/c seen to crash, hit by HAA or LAA 2 destroyed by RAF & 2 or 3 others were possibles for AA guns
24th Feb						
2340-2352	108	135	93	Not recorded	2	A/c turned by barrages
26th Feb						
0400-0500	525	458	469	4,362	12	1 a/c hit by 24 Bty and believed destroyed 1 other also believed destroyed by HAA
27th Feb						
0355-0500	383	188	398	2,981	8+	Bombs dropped short
28th Feb						
0325-0530	285	378	308	3,141	4+	Raids lasted until 0530
1855-1928	207	*	225	1,829	6	Determined attacks; bombs outside hbr
2220-2250	22	*	12	194	1	Minelayer (claimed by RAF)
2nd Mar						
1825-1908	715	*	716	5,411	10+	1 a/c destroyed & 1 prob to HAA & LAA
3rd Mar						
1829	16	—	—	Not recorded	1+	No warning from GOR
4th Mar						
0558	16	—	—	Not recorded	Not known	Engaged on sound of a/c
7th Mar						
1845-1909	407	354	342	2,723	17+	1 a/c destroyed & 1 prob to AA 40 bombs dropped; some damage
8th Mar						
1826-1846	28	32	32	Not recorded	2	A/c turned away by barrage
11th Mar						
1945-2008	342	271	257	1,967	5+ (26 Bty stated 12)	No damage or casualties

Date and Time	Rounds fired by 9th (Londonderry) HAA Regiment 24 Bty	25 Bty	26 Bty	Total HAA Rounds fired (all units)	No of a/c involved	Comments
15th Mar[1] 2121-2327	139	170	186	1,189	2 + 3	No warning. Bombs fell in defended area. No damage to shipping
16th Mar 2050-2130	766	715	602	5,501	10+	4 a/c destroyed by AA guns
19th Mar 1847-1859	221	224	116	1,309	15+	6 a/c destroyed by AA; 1 probable & 1 damaged. Ships in harbour hit and destroyed.
20th Mar 0150	12	—	—	Not recorded	NK	Fired on sound of a/c
0308	12	—	—	Not recorded	NK	Fired on sound of a/c
1324	—	—	—	None recorded	1 x Ju88	Recce
24th Mar 1232	14	—	—	Not recorded	1 x Ju88	A/c dived from 23,200 feet to 17,000 on being engaged
29th Mar 1218	33	—	20	Not recorded	1	A/c dived on being engaged. 1 rd seen to burst under tail
2nd Apr 0915	—	—	—	Nil	1 x Ju88	Intercepted by RAF
4th Apr 0657	3	2	—	Not recorded	1 x Ju88	
5th Apr 0725	9	13	16	Not recorded	1 x Ju88	
12th Apr 1220	27	14	16	Not recorded	1 x Ju86	Damaged by HAA
13th Apr 1930-1953	407	307	276	2,653	12+	"very half-hearted"
15th Apr 1030	8	—	—	Not recorded	1 x Me210	A/c turned away
26th Apr 0237	24	24	—	262	1	Nightfighter intercepted

* 25 Bty's War Diary does not note these raids but it is unlikely that the Battery was not involved as ex-members of the Battery are adamant that they were involved in all the major actions.

[1] The totals for each of the three Batteries are believed to include those for a raid by 3 aircraft at 2307. The RHQ War Diary shows only 44 HAA rounds fired at 2121.

Abbreviations:
A/c or a/c—aircraft   rds—rounds   tnk—type not known   dam—damaged   prob—probable   GOR—Gun Operations Room

APPENDIX 6

ROLL OF HONOUR: THEIR NAME LIVETH FOR EVERMORE

The following members, or ex-members, of 9th (Londonderry) Heavy Anti-Aircraft Regiment, Royal Artillery died on active service as the result of enemy action or (in a very few cases) from disease or accident.

G Millar	26 HAA Battery
J McMeekin	26 HAA Battery
H Gamble	5 LAA Battery*
W Graham	5 LAA Battery*
T Porter	25 HAA Battery
J J McClean	24 HAA Battery
M McCartney	24 HAA Battery*
J McNeilly	5 LAA Battery
S Graham	5 LAA Battery
J Dalzell	5 LAA Battery
J Meredith	5 LAA Battery
C Boyes	5 LAA Battery
P Robson	5 LAA Battery
B Martin	5 LAA Battery
E Shiels	5 LAA Battery
J Gibson	5 LAA Battery
A Davidson	5 LAA Battery
D McClinton	5 LAA Battery
Rayner	5 LAA Battery
R Bailey	5 LAA Battery
A Gage	5 LAA Battery
R Cromie	5 LAA Battery
J Miskelly	5 LAA Battery
E Flynn	5 LAA Battery
H Johnston	5 LAA Battery
P Toner	6 LAA Battery
R Dunlop	6 LAA Battery
E Rutherford	26 HAA Battery/MAC Troop
N Murray	25 HAA Battery/MAC Troop
T McCarter	25 HAA Battery/MAC Troop
J Gilmour	25 HAA Battery
D Mark	25 HAA Battery
A Evans	25 HAA Battery
R Hamilton	25 HAA Battery
W Peoples	25 HAA Battery
C Kharasi	25 HAA Battery

S Mosesetheli	25 HAA Battery
R Khoamyane	25 HAA Battery
M Nena	25 HAA Battery
N Tsoasane	25 HAA Battery
T Lepota	25 HAA Battery
L Hlakane	25 HAA Battery
S Chele	25 HAA Battery
S Ntsoebe	25 HAA Battery
A Semane	25 HAA Battery
S Motuba	24 HAA Battery
J Randall	25 HAA Battery*
Wilson	25 HAA Battery*
W E Erwin	24 HAA Battery*
W McBlain	5 LAA Battery**
T McCutcheon	5 LAA Battery**
J Shanks	5 LAA Battery*
J Brown	5 LAA Battery*
J McCartney	24 HAA Battery
W McGall	26 HAA Battery
J McNeill	26 HAA Battery
Mofosi	26 HAA Battery
J Routledge	25 HAA Battery
A McKeown	6 LAA Battery
R Coleman	26 HAA Battery*
T Lennox	24 HAA Battery
J Kelly	24 HAA Battery*
T King	24 HAA Battery*
J Shields	25 HAA Battery*
J Stevenson	25 HAA Battery

SPLENDID YOU PASSED, THE GREAT SURRENDER MADE,
INTO THE LIGHT THAT NEVERMORE SHALL FADE

Notes: Names marked thus * indicate that the person named was no longer serving with the Regiment or one of its Batteries at the time of death.

Names marked thus ** indicate persons who died as Prisoners-of-War.

Whilst every effort has been made to ensure the accuracy of this list it is appreciated that errors may have occurred and that some names may have been omitted.

BIBLIOGRAPHY

UNPUBLISHED SOURCES

WAR DIARIES (Public Record Office, Ruskin Avenue, Richmond, Surrey)

(i) From the series WO 169 (Middle East Forces):

Regimental HQ, 9th (Londonderry) HAA Regiment — a total of five Diaries covering the period November 1939 to December 1943.

24 HAA Battery — a total of four Diaries covering the period from December 1939 to December 1943.

25 HAA Battery — a total of four Diaries covering the period from June 1940 to December 1943.

26 HAA Battery — a total of seven Diaries covering the period from November 1939 to December 1943.

5 LAA Battery — two Diaries covering the period from August 1939 to December 1941. No Diary has survived for 1942.

6 LAA Battery — a total of four Diaries covering the period from August 1940 to November 1943.

20 HAA Battery — one Diary covering the period August 1939 to September 1940.

5 HAA Battery, Royal Malta Artillery — two Diaries covering the period May 1940 to May 1941.

Other Diaries consulted from this series were those of HQ, Royal Artillery, Alexandria 1939-40; HQ, Anti-Aircraft, Alexandria, 1941 and 1942; Commander, Royal Artillery (CRA), Western Desert Force 1940; Matruh Fortress 1940-41; Gun Operations Room, Alexandria, 1941; Brigadier, Royal Artillery (BRA), AA, Middle East Forces 1940-41; 187 HAA Bty 1941; 2nd Anti-Aircraft Brigade Tripoli 1943; 12th Anti-Aircraft Brigade Italy 1943; ME Counter-Battery Pool 1943.

(ii) From the series WO 170 (Central Mediterranean Forces):

Regimental HQ 9th (Londonderry) HAA Regiment from January to August 1944.

24 HAA Battery from January to August 1944.

25 HAA Battery from January to August 1944.

26 HAA Battery from January to August 1944.

66 AA Brigade from January to August 1944.

(iii) From the series WO 166 (UK Home Forces):

Regimental HQ, 9th (Londonderry) HAA Regiment — two Diaries covering the period September 1944 to November 1945.

25 HAA Battery — one Diary for September to October 1944.
26 HAA Battery — one Diary for September to October 1939 and one Diary for September to October 1944.
Other Diaries consulted from this series include 41st AA Brigade and 5 AA Group, 1945, and Anti-Aircraft Command HQ, 1945 and 1946.

(iv) From the series WO 171 (Allied Expeditionary Force, NW Europe):
27 LAA Regiment — two Diaries covering the period January 1944 to October 1945 (6 LAA Bty was regimented in 27 LAA Regt during this period; no Battery War Diary can be traced).

THE McFARLAND PAPERS (Public Record Office of Northern Ireland, Balmoral Avenue, Belfast), by permission of Sir John McFarland, Bt TD

(i) 24 HAA Battery History — contemporary notes covering the period 1939 to 1946.

(ii) 25 HAA Battery History — contemporary notes covering the period 1939 to 1944 (August).

(iii) Correspondence between the War Office and Sir Basil A T McFarland Bt on the Honorary Colonelcy of 9th (Londonderry) HAA Regiment 1946.

(iv) Correspondence between Sir Basil A T McFarland Bt and the Commanding Officer of 9th (later 60th) (Londonderry) HAA Regiment 1946-47.

(v) Correspondence between the Co Londonderry TA Association and 60th (Londonderry) HAA Regiment regarding the "parenting" by that unit of 246 (Derry) (Mixed) HAA Regiment, RA (TA) 1947.

PERSONAL ACCOUNTS, JOURNALS AND DIARIES

(i) "TWO-FIVE — Experiences of a HAA Battery" by an officer of the Battery (believed to have been John Kershaw) 1945. By kind permission of Sir John McFarland Bt TD.

(ii) Private Diaries (contemporary) Mr John J Ormsby.

(iii) "FROM BALLYMENA TO NAPLES" — a journal by Mr Tom Moore (ex 26 HAA Battery) 1946.

(iv) "FROM AGAMI TO THE ARNO" — an account of F Section, 25 HAA Battery by William Sherrard.

ROYAL ARTILLERY INSTITUTION LIBRARY
"THE DEVELOPMENT OF ARTILLERY TACTICS AND EQUIPMENT" by Brigadier Pemberton 1951.

PUBLISHED SOURCES

THE SECOND WORLD WAR — Books IV, V, VI, VII, VIII, IX, X and XI by Winston S Churchill, published by Cassell & Co (1949).

ACK-ACK by General Sir Frederick Pile, published by Harrap (1949).

GUNNERS AT WAR by Brigadier Peter Mead, published by Ian Allan (1982).

AND WE SHALL SHOCK THEM by General Sir David Fraser, published by Hodder & Stoughton (1983).

BRITISH AND AMERICAN ARTILLERY OF WORLD WAR TWO by Ian V Hogg, published by Arms and Armour Press (1978).

WAR IN THE DESERT — EIGHTH ARMY AT EL ALAMEIN by James Lucas, published by Arms and Armour Press (1982).

THE BRITISH ARMY IN THE TWENTIETH CENTURY by Ian V Hogg, published by Ian Allan (1985).

NORTHERN IRELAND IN THE SECOND WORLD WAR by Professor J W Blake, published by HMSO (Belfast) (1956).

2nd NEW ZEALAND DIVISIONAL ARTILLERY by W E Murphy (Official History of New Zealand in the Second World War 1939-45) published by Historical Publications Branch, Department of Internal Affairs, Wellington (1966).

HISTORY OF THE SECOND WORLD WAR by Liddell Hart, published by Pan (1973).

THE MILITARY HISTORY OF WORLD WAR TWO edited by Barrie Pitt, published by Guild Publishing (1986).

THE ARMY AND AVIATION by Bruce Robertson, published by Robert Hale (1981).

WITH ROMMEL IN THE DESERT by Heinz Werner Schmidt, published by Harrap (1951).

ROMMEL — BATTLES AND CAMPAIGNS by Kenneth Macksey, published by Arms and Armour Press (1979).

THE TRAIL OF THE FOX — THE LIFE OF FIELD MARSHAL ERWIN ROMMEL by David Irving, published by Weidenfeld and Heinemann (1983).

OPERATION AVALANCHE by Des Hickey and Gus Smith, published by Heinemann (1983).

THE LONG MARCH ON ROME by Charles Whiting, published by Century Hutchinson (1987).

ROME 1944 by Raleigh Trevelyan, published by Secker and Warburg (1981).

CALCULATED RISK by General Mark Clark (memoirs) published by Harper, New York (1950).

LUFTWAFFE by Alfred Price, published by Pan/Ballentine (1973).

BRITAIN UNDER FIRE by Charles Whiting, published by Century Hutchinson (1986).

MONTY — THE MAKING OF A GENERAL by Nigel Hamilton, published by Hamish Hamilton (1981).

MONTY — MASTER OF THE BATTLEFIELD by Nigel Hamilton, published by Hamish Hamilton (1983).
THE MEMOIRS OF FIELD MARSHAL THE VISCOUNT MONTGOMERY OF ALAMEIN, KG published by Collins, London (1958).
CHURCHILL AND THE IRISH MARSHALS by Patrick Marrinan, published by Pretani Press, Belfast (1986).
AN IMPROVISED WAR (The Abyssinian Campaign of 1940-41) by Michael Glover, published by Leo Cooper (1987).
SOLDIERS — A HISTORY OF MEN IN BATTLE by John Keegan and Richard Holmes, published by Hamish Hamilton (1985).
DECISION IN NORMANDY by Carlo D'Este, published by Collins (1983).
A GUARDS' GENERAL — The memoirs of Major General Sir Allan Adair Bt, GCVO, CB, DSO, MC, JP, DL, edited by Oliver Lindsay F R Hist S, published by Hamish Hamilton.
AN ACCOUNT OF THE TERRITORIALS IN NORTHERN IRELAND 1947-78 compiled by Colonel I B Gailey, Colonel W F Gillespie and Lt Colonel J Hasset, published by the TAVR Association, NI (1979).
THE MILITIA ARTILLERY 1853-1909 by Norman E H Litchfield, published by N E H Litchfield (1987).

POSTSCRIPT: 1947 TO 1987

The post-war forces underwent a reduction in their strength that saw the army fall from almost three million men in July 1945 to some 750,000 two years later. Demobilisation had been swift and the citizen-soldiers had returned to their civilian lives and jobs.

When that contraction had been achieved in the Regular Army the Territorial Army was reformed to provide back-up for its full-time partner. This reborn TA was extended immediately to include Northern Ireland, which had been excluded from the TA structure in 1920 as indeed had been all of Ireland from the original Territorial Force of 1908. The new TA was to include from the outset a significant number of Gunner units in Northern Ireland — in itself this was a tribute to the achievements of the men who had provided the units of 3rd (Ulster) AA Brigade in 1939.

Before the new regiments were created however the Royal Regiment of Artillery itself went through what was probably the greatest transformation in its history when, with the exception of the Royal Horse Artillery, all of its Batteries and Regiments were placed on to a single roll. This eliminated the possibility of confusion between two, or even more, batteries with the same number but a different function as could have happened with the old system whereby the regiments and batteries were numbered within each role. Thus for example there was a 9th Field Regiment as well as a 9th HAA Regiment and with field, medium, heavy, super-heavy, anti-tank, light AA, heavy AA, coastal, searchlight and rocket batteries and regiments there was considerable scope for confusion. The simple numerical system which was introduced was not quite so simple and straightforward as it appeared, as some of the most senior batteries found themselves with higher numbers than some of their juniors.

The re-organisation saw 9th HAA Regiment become 60th HAA Regiment while 24, 25 and 26 HAA Batteries became respectively 151, 155 and 166 HAA Batteries. They had also absorbed on the way the heritage of 62, 117 and 84 Field Batteries respectively! With these designations the Regiment soldiered on until its final disbandment on December 15th 1955. 151 Bty became independent as an AA Fire Control Battery until it too was disbanded on January 1st 1962. 155 Bty was placed in suspended animation in 1955 and disbanded on the same day as 151. 166 survived for a time in 95th AO Regiment. Although there are still batteries bearing the numbers 24 and 25 in 1988 neither have any connection with the batteries of 9th Regiment — today's 24 is 24 (Irish) Battery and is a training unit in the Royal Artillery.

Because of the new system the TA gunner units received very high numbers indeed and when the designations of the new Northern Ireland volunteer artillery regiments were announced in 1947 this proved to be the case. There were seven gunner regiments in all and, but for a field regiment and a coastal regiment, all were dedicated to the AA role. The new AA units were:

245 (Belfast) (Mixed) HAA Regiment, Royal Artillery (TA)
246 (Derry) (Mixed) HAA Regiment, Royal Artillery (TA)
247 (Ulster) Light AA/Searchlight Regiment, Royal Artillery (TA)
248 (Ulster) Light AA Regiment, Royal Artillery (TA)
502 (Ulster) (Mixed) HAA Regiment, Royal Artillery (TA)

246 (Derry) (Mixed) HAA Regiment was "parented" by 60th (Londonderry) HAA Regiment, a War Office linking which was designed to save that department the expense of providing the initial cash for the Regimental Funds of each of the new units. Instead the "parent" unit had to provide this money for its linked junior and, in spite of being short of funds itself, 60th HAA Regiment had to make a payment of £240 to the coffers of 246 HAA Regiment. Fortunately this did not cause any friction between the two regiments.

Other close connections with wartime days were to be found in the personnel of the new regiments. In Londonderry the first Commanding Officer of 246 HAA Regiment was Lt Colonel J H (Jack) McVeigh who had been commissioned into 25 HAA Battery in 1939 and had later commanded 26 Battery. The Regiment's Honorary Colonel was of course Sir Basil McFarland while the Colonelcy of 248 LAA Regiment went to Colonel Sam Haughton, ERD, MP, DL who had commanded 26 Battery also. 248 Regiment was based at Coleraine with a battery at Ballymena and in 1953 it received a new Commanding Officer — Lt Colonel D J Christie, CBE, ERD who had commanded 6 LAA Battery in North Africa and North-West Europe.

All of the Anti-Aircraft regiments were brigaded in 51st (Ulster) AA Brigade (TA), which was strengthened in July 1948 by the addition of a Regular HAA Regiment which moved into Palace Barracks near Belfast. This regiment which had moved from eastern England was none other than 60th HAA Regiment — the old 9th in new guise. The move fulfilled the aim of the Commanding Officer, Lt Colonel Jim Longfield, to ensure that the connection of the Regiment with Northern Ireland should remain as close as possible — he had in fact wanted the Regiment to move to Londonderry but Belfast was as close as he could manage.

However the days of the heavy manned AA gun were now numbered. The advent of the A-Bomb made it possible for a single bomber to cause untold destruction with one weapon. Thus it would no longer be sufficient for the defenders to be able to turn away or shoot down most of the attackers; not even one aircraft could in future be allowed near the target. As a result the development of the guided surface-to-air missile went on apace to take over the role of the heavy gun. As missiles were perceived as being aircraft of a sort rather than shells the Royal Air Force laid claim to their control and in the early 1950s the junior service was given full responsibility for such missiles whenever they would enter service. This policy was later modified to allow the Royal Artillery to operate similar weapons for the protection of the field forces and a missile for gunner use was also developed.

The introduction of the SAMs brought the days of the 3·7 inch gun to a close. Bloodhound SAMs were introduced into RAF service and their sole task was the defence of Bomber Command Bases — groups of sixteen Bloodhounds were allocated

Lt Colonel D J Christie, CBE, ERD, DL, JP in the Dress Uniform of Commanding Officer of 248 (Ulster) Light AA Regiment, RA (TA) 1953.
(Photo: Lt Colonel Christie)

General Sir Thomas Morony, the Master Gunner, tries out a Blowpipe Training Unit while on a visit to 206 (Coleraine) Air Defence Btn.
(Photo: 206 (Coleraine) AD Btn with the permission of the Master Gunner)

to RAF squadrons from 1958 onwards. In 1960 36 Heavy Air Defence Regiment, RA introduced the Thunderbird SAM to service but before either missile was operational the Army's Anti-Aircraft Command, which had operated under the RAF during the war, had been disbanded. In its place there remained only three regiments with 3·7 inch guns forming 7 AA Group, RA.

With the demise of AA Command came the disbandment of most of the heavy regiments. In March 1955 246 (Derry) (Mixed) HAA Regiment (TA) ceased to exist; in July 1955 60th HAA Regiment went into suspended animation at Palace Barracks, and on December 15th the disbandment process was complete and the old 9th finally came to an end as a HAA unit after sixteen years of service in war and peace.

However there was still a role for LAA guns in direct defence of ground troops, and so the decision was taken to amalgamate all of the Territorial regiments of 51st (Ulster) AA Brigade into one Light AA regiment. The new unit, at first to be designated 434 LAA Regiment, was formed as 245 (Ulster) LAA Regiment and went off to its annual camp that August with a fair proportion of former members of 246 (and even some wartime members of 9th). As part of 41 Army Group, RA the Regiment had its HQ and two of its Batteries in Belfast with Troops of Q Battery at Coleraine, Londonderry (E Tp) and Ballymena (F Tp). The protests of the Honorary Colonel, Sir Basil McFarland, that a Battery HQ should have been sited in Londonderry were to no avail and Sir Basil's worst fears were realised in 1961 when, in a further re-organisation, both Londonderry and Ballymena lost their gunners after twenty-one years' direct affiliation with the Royal Regiment in the air defence role. For Londonderry it was the end of an even longer connection with the gunners stretching back to the original Londonderry Artillery Militia of 1855.

245 LAA Regiment was redesignated a Light Air Defence (LAD) Regiment in 1964, but in 1967, during further restructuring of the reserve forces, it was amalgamated with 445 (Lowland) LAD Regiment to form 102 (Ulster and Scottish) LAD Regiment, RA (V). The only gunner connections left in the province were significantly the homes of the old LAA batteries — 5 and 6 — at Newtownards and Coleraine as RHQ and HQ Battery were based at the former with 206 (Ulster) Battery at the latter. With the withdrawal of the Bofors 40 mm gun in 1978 and the adoption of portable guided missiles the Regiment was restyled an Air Defence (AD) Regiment.

Further change took place in 1986 when the Scottish elements of the Regiment were hived off to form a new regiment and 102 became 102 (Ulster) AD Regiment with 206 AD Battery becoming 206 (Coleraine) AD Battery, RA (V).

102 (Ulster) AD Regiment today displays a level of activity not found in 1939 when the gunners of 9th Regiment found themselves thrown into war with little preparation. As part of an expanded British Army of the Rhine in the event of future hostilities 102 (Ulster) Regiment is a highly professional part of the deterrent which will hopefully persuade any would-be aggressor that there could be no winners in another world war. As part of that Regiment 206 (Coleraine) AD Battery carry on the traditions so proudly upheld by their predecessors, not least of who were the "Derry Boys" of 9th (Londonderry) HAA Regiment, Royal Artillery.